HOW TO WORK WITH
CONCRETE AND MASONRY

How to Work With

CONCRETE

and MASONRY

by Darrell Huff

Illustrations by Carolyn and Gerald Kinsey

Popular Science Publishing Co., Inc.   •   Harper & Row
A Times Mirror Subsidiary   New York, London

CONTENTS

Introduction

THE THINGS you can do with concrete and stone in improving, remodeling or even building your home or vacation home are just about limitless. The versatility of masonry, and its durability when the work has been properly done, are phenomenal.

One of the rewarding things about working with rock and cement in any of their many forms is that you get a lot for a little money. Another is the great control you can exercise over the result. All masonry is what *you* make it.

A good deal of what you find in this book comes out of my own experience in building a number of houses and vacation cabins and in adding patios and porches, fireplaces and barbecues, walks and driveways and pools of one kind or another to make the dwellings more enjoyable to live in. A camera was kept at hand a large part of the time, and that's how nearly half of the more than 200 photographs in this book were made.

No one gets very far into the lore of concrete work without turning for help to the Portland Cement Association, 33 West Grand Avenue, Chicago, Illinois 60610. I want to acknowledge with gratitude the very considerable help they have given me in supplying information about their product, as well as illustrations. Thanks are also due to the National Concrete Masonry Association, the Wire Reinforcement Institute, Kaiser Cement & Gypsum Corporation, and the California Redwood Association.

I want to thank also the editors of *Popular Science Monthly,* and Robert P. Stevenson in particular, for their encouragement and guidance in the original accomplishment of many of the projects that provided a solid background for this book.

Thanks are owed above all to my wife and children, and to our many friends and unwary weekend guests, for their labors in mixing and troweling. A daughter, formerly one of these captive assistants, is now half of the team of engineer and artist, Gerald and Carolyn Kinsey, who produced the sketches and diagrams that I hope will help to make this book useful to you.

PART I
WORKING WITH CONCRETE

CHAPTER ONE

THE BASICS OF
GOOD CONCRETE WORK

FOR THE beginner, this chapter is intended as an introduction to do-it-yourself concrete work. The Old Hand can use it, too, as a refresher in basic techniques. It may remind him that he is doing some things the hard way, and it may help him to make his jobs easier, faster, and better.

Concrete is one of the oldest and most versatile of plastics. It is made of gravel and sand held together by a paste of portland* cement and water. You can buy these ingredients and make your own concrete, or you can buy concrete ready-made. In the first case you will order a load of rocks and sand—separately, or mixed together to form what is called concrete aggregate—and obtain cement in 94-pound sacks from a dealer in building materials. Ready-made concrete comes in two varieties: mixed with water and ready to place; or as dry-mix, needing only water to complete it.

You will probably have occasion to use several of the various methods of obtaining concrete. No one of them is the perfect answer to every problem. This being the case, let's begin by considering the one that involves the least work on your part and work our way up.

Naturally, the ideal way to get concrete is to buy it already mixed and delivered to your home and poured into your forms or your wheelbarrow. This is exactly what you are getting when you buy transit-mixed concrete; and the funny thing is that in most localities this is absolutely the cheapest way to buy concrete as well as the easiest. And you get the best possible concrete this way, too.

So it naturally follows that the first rule of concrete work is this: use transit-mix if you can. A later chapter is devoted to tips on ordering and placing this kind of concrete.

For some around-the-home jobs you can't use transit-mix. Either the place you want it is unreachable by heavy truck, or the quantity you need is too small to make its use feasible. Some companies will not deliver amounts under one cubic yard; those that will do so charge a high price for the service.

---

*Nothing to do with either Oregon or Maine. Common cement takes this name from its resemblance to stone quarried on the English Isle of Portland.

This brings us to your second choice: you-haul concrete. This, too, is sold to you already properly proportioned and mixed. But you haul it to the job yourself. You do this with a rented trailer supplied by the dealer. Some dealers supply a pickup truck instead. Another chapter in this book deals with using this type of concrete.

If you-haul concrete won't work for you, or if it is not available where you live, you must mix your own. You can do this in a revolving-barrel mixer powered by either an electric motor or a small gasoline engine. Into this barrel you pour water and shovel cement, sand, and gravel. Then you tilt it to dump the mixed concrete into a wheelbarrow. Instead you may prefer to use a combination wheelbarrow and mixer. With this device you shovel in the makings in just the same way. But instead of dumping the mixed mud into a vehicle, you just pick up the arms of the wheelbarrow-mixer and trundle it to where the concrete is to be dumped. A later chapter tells in more detail about this and other mixers and how to use them.

That same chapter also covers your next choice—mixing by hand. Doing this, whether in a large box, in a wheelbarrow, or on a slab of concrete, is pretty laborious for any but the smallest jobs. I happen to know a man who built a large stone-and-concrete house, mixing every ounce of the concrete (every ton would be a better way to put it) by hand. This proves what can be done. But all the same, I wouldn't recommend it.

There is still another way to obtain concrete for your projects. Buy it in the form of dry-mix. This is a mixture of sand and gravel and cement, thoroughly dried and properly proportioned. It is put into sacks and sold in lumberyards, patio supply shops, and hardware stores. As with so many products these days, all you have to do with it is add water. The drawback to dry-mix is that it necessarily is comparatively expensive. You'll want it only for small jobs. For these it—like so many modern convenience products—is well worth the cost.

In the case of mortar, for use in laying up stone, brick, or block, your choices are more limited. You can either mix your own—from sand, cement, and water with the possible addition of lime—or you can buy it in sacks as dry-mix mortar. You'll normally do the former for large jobs, the latter for small ones.

Estimating the amount of concrete you will need for a job is not especially complicated. Usually you will measure the length and breadth in feet and the thickness in a fraction of a foot. Multiply these three figures together to find how many cubic feet you will need.

Although there are 27 cubic feet in a cubic yard, it is simpler to use 25. This automatically gives you about an 8 percent allowance for waste and for inaccuracies in measuring.

Suppose you are pouring a sidewalk 3 feet wide and 20 feet long with usual average thickness of about 4 inches—which is one-third of a foot. One-third of 3 times 20 is 20. You'll need 20 cubic feet, or about four-fifths of a cubic yard of concrete.

If you are building a retaining wall that is to be 8 inches thick, consider this as eight-twelfths, or two-thirds, of a foot. If the wall is to be 2 feet high and 18 feet long, take two-thirds of 18 times 2—which is 24. Mix or buy a cubic yard of concrete and you should have just a little bit to spare.

It might be helpful right now just to list the steps involved in doing a typical concrete job. All these steps are covered in more detail in later chapters, where there are also photographs to help make things clear.

1. Excavate and prepare the base for the concrete.
2. Build the forms and brace them solidly.
3. Place reinforcing steel if you are using any—as you usually should do.
4. Make arrangements for one or more helpers unless it's a one-man job.
5. Order your concrete or ingredients.
6. Assemble tools—usually rake, shovel, floats, trowel. But don't forget gloves and possibly boots.
7. If necessary, wet down the area so concrete won't dry too fast. Or place waterproofing membrane if needed.
8. Mix the concrete or dump it from the transit-mix truck. Move it to the site and spread it with rake and shovel, spading and compacting.
9. Strike off level, usually with a 2-by-4.
10. Smooth the surface with a wood float.
11. Complete finishing with steel trowel or with a broom, depending on finish wanted.
12. Keep concrete moist for at least three days for proper hardening.

CHAPTER TWO

HOW TO BUILD FORMS

WE MIGHT as well begin by facing the fact that form building is one of the most frustrating kinds of carpentry. When you build a form you have to do it very accurately—level, plumb, square, and all that—or your final product in imperishable concrete will be askew. You have to nail your form strongly and brace it solidly or it may collapse in mid-pour. Having done all this, you must then attack the form again a few hours or a few days later and tear it all down.

Nobody has found a way around this dilemma yet, so you can only accept it. Even the people who build the giant concrete over- and underpasses for highway intersections have to do it this way.

For economy of both money and effort, there are several points to keep in mind when building forms.

1. Try to use large and heavy materials that you already have on hand for other purposes—and can then clean up and reuse. Suppose you're building a vacation cabin that will require some large ceiling beams. They will be rigid enough to be used as outside forms for the concrete-slab floor with only a few stakes required.

Or suppose you need to pour a retaining wall as part of a building project. If the subfloor of the building is to be of the usual 2-by-6 tongue-and-groove lumber, use it first for the form for that wall.

An excellent material for building wall forms is heavy plywood. This is also suitable for subfloors. So use it first for form work, then salvage it for the subfloor job.

2. Brace your forms against two things: bulging and movement. You don't want the weight of the concrete to distort the forms. But it is just as important to make sure that a whole form won't be pushed out of place. Generally you can accomplish both these purposes by making liberal use of 2-by-4 lumber as stakes and as whalers—horizontal pieces that keep the form from bulging. Since there are so many uses for 2-by-4's in almost any building project, you can salvage your stakes and whalers with little waste.

3. In designing landscaping, look for ways to make the forms a permanent part of the project. When you pour a patio with redwood grids, as described in a later chapter, you have no forms to tear out later. The grids become a permanent and decorative part of the patio.

A similar method can be used to make unusually attractive steps in a

4

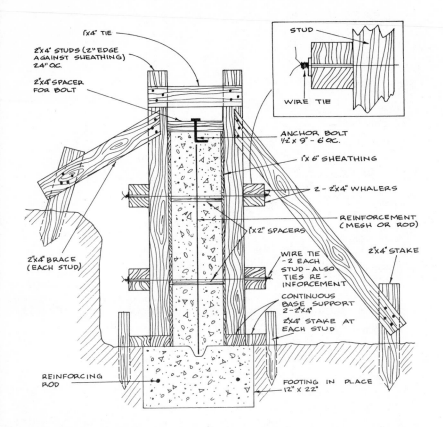

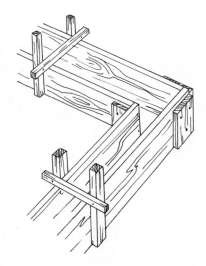

Conditions vary so much that each form you build will differ a little from the last one. But you can easily adapt the basic arrangement shown in this diagram to form up for almost any wall you may have occasion to construct.

This portion of a form for a lawn curbing illustrates some additional principles. Stakes are driven down or sawed off flush with the top of the form where possible, so they won't interfere with striking off the concrete level. Wooden ties between stakes are used to keep the sides of the form from spreading when the concrete is poured. If the top of the wall or curbing is to be finished uniformly, it is best to place these ties at least an inch above the top of the form so they won't be in the way. A curb wall should be carried down to solid bearing. If the earth is firm the sides of the excavation will serve as forms for the part of the wall that is below grade. If not, forms will have to be carried to the bottom of the excavation. A curbing like this should be at least 4 inches thick.

When building forms for a narrow sidewalk, you can use a few temporary posts to establish the grade. Lightly nail, or clamp, one edge 2-by-4 to a stake and then use a level to set the other edge form. A small level placed on a board will serve in the absence of a long mason's level like this one. Permanent stakes, usually of 1-by-4, can then be driven outside the forms and nailed to them.

terraced garden. Build the steps in the shape of redwood boxes—rough 2-by-6 redwood is a good choice—and then just fill the boxes with concrete.

4. Build forms so they will be easy to take down later. A most helpful item to this end is the special kind of nail known variously as a duplex, scaffold, or form nail. It has two heads and so is easily grasped for removal. A wrecking bar is an inexpensive tool with which you can pull such nails much more easily than with the claws of a hammer.

5. Prepare in advance for emergencies that may arise during pouring. Keep some extra stakes and bits of lumber for bracing, so that you can nail them into place if your forms threaten to give way after you have begun to

Forms for a combination ribbon driveway and sidewalk look like this. The pouring and leveling process will be much simpler if the stakes are first cut off to form level as has been done with those in the foreground. Welded wire fabric reinforcement has been cut and lapped. It will be hooked and lifted to the midpoint of the slab as the concrete is placed.

Forms for a straight driveway are very simple—just 2-by-4s for edge forms, with 1-by-4 stakes every few feet. Reinforcing fabric, seen ready to be hooked into place, is especially important in a driveway subject to great temperature changes and occasional heavy loads.

pour concrete. This is most likely to happen when you are pouring a fairly high wall. With forms for this purpose another good piece of emergency equipment is a few woodworking clamps. You can use these not only to stop bulging of forms but even to pull a form back together after it has started to bulge.

6. Design your forms so they will be the least possible nuisance during the concrete work. This means you should avoid placing stakes and braces where they will be in the way of a concrete truck or the wheelbarrow. Where you will have to pull a strike-off board along the top of the form—as with a sidewalk—saw off all stakes level with the top of the form. Otherwise you'll have to work around each stake as you strike off the concrete to make it level.

BUILDING A BASIC WOOD FORM. The drawing shows these points applied to building a basic wood form for pouring a concrete wall. To build a form such as this, you will normally begin by laying the continuous base supports of 2-by-4 lumber. A second 2-by-4 against each is wedged between it and a series of short stakes.

You can then erect the main supporting posts, or studs. These are 2-by-4's

7

Economical reuse of form materials is shown in this photograph of the author's house under construction. The edge forms of 4-by-8 lumber will be cleaned and used as joists for the second story. Wall forms use 1⅛-inch plywood which will become subflooring. Rocks are laid ready to be dropped in during the pour, to give the effect seen in the finished wall at the right. The 1-by-12 lumber in the foreground is used above each piece of plywood to carry the wall to the 5-foot height wanted.

that are 6 inches or more higher than your wall is to be. Place these edgewise, just as you would in building a stud wall of a house, and toenail them to the base support. Use doubleheaded nails for easy removal.

The actual form boards are nailed to these studs. If you use 1-inch lumber, or plywood, you will facenail through the boards and into the studs. To take the forms down later, you will pull posts and boards off as a unit and then remove the nails. If you use 2-inch lumber, such as Douglas fir or pine subfloor tongue-and-groove stock, you can nail from the outside by going through the corners of the 2-by-4 posts and into the t&g material. Use a C-clamp to hold the lumber together while nailing.

Whalers are 2-by-4's placed horizontally against the posts to help keep the wall straight. They can be held by duplex nails driven at an angle.

To hold the two sides of the wall the proper distance apart at the top, tie each pair of studs together with a short length of 1-by-4 lumber.

A high wall will also need to be held both apart and together at intervals between top and bottom. Little spacers of 1-by-2, their length being the thickness of the wall, will hold the two sides of the form apart. You will pull these out as you pour. Hold the sides of the form together by tying with wire at each point where a whaler crosses a stud. This is also good placement for the spacers.

With all this done, your form should be strong and rigid and of uniform thickness. To make it plumb and hold it that way, nail a diagonal brace to each stud near the top. After making the wall straight up and down, and checking with a level, secure each diagonal brace to a short stake driven into the ground.

Because the circumstances vary, you may never build a form in exactly this way. But by combining the basic method described here and shown in the diagram with the six tips given earlier, you can build a suitable form for almost any purpose. Details of specific forms for other purposes are covered in later chapters.

Form work can be simplified, as shown here, for low walls where precision is not important. Since this wall will be covered when the two-level floor slab is poured, it need not be perfectly straight. Otherwise whalers would be needed to strengthen the forms and keep them straight during the pour.

CHAPTER THREE

HOW TO
REINFORCE CONCRETE

THERE ARE three good reasons for using steel reinforcing on most of your concrete jobs.

With slabs, sidewalks, driveways, and stepping stones, welded wire fabric will spread the stresses and reduce the likelihood of cracking. If fine cracks do appear, the wire fabric will still hold the concrete together. In this way it will add strength at less cost and with less labor than if you simply added more concrete to make a thicker slab.

Where concrete is unsupported, as with walls, fireplaces, and foundations, reinforcing rod will actually convert your concrete into a new and far stronger material. When reinforced with rod, it becomes ferroconcrete. Such things as cantilevered steps or hearths are hardly feasible without reinforcing rod.

A third reason for learning to use reinforcing is that on some jobs you *must.* Such structures as foundations and fireplaces for houses covered by building codes must usually be reinforced in specified ways. Unless your plans call for using steel rod and you have the required amount either placed or ready at hand, the building inspector will ordinarily not give you an okay to pour.

WELDED-WIRE FABRIC. This material looks like heavy fencing wire. It usually comes in long rolls 5 or 6 feet wide. You can have as many running feet of it as you need cut from a roll at your lumberyard or building-supply dealer. Because it's a nuisance to unroll this stuff and cut it off for you, the price is usually a good bit lower when you buy it by the full roll. So it is an economy to plan ahead and buy a whole roll if you will eventually use it up.

The most widely available welded-wire fabric is the kind designated as 6x6-10/10. This means the wires are spaced 6 inches apart each way and that they are 10-gauge wire. As the chart shows, it is better to use heavier or more closely spaced wire for some jobs.

Easiest tools to cut this stuff with are wire cutters or small bolt nippers. The lighter gauges can be handled fairly well with a heavy cutting pliers. If you don't have too much of it to cut, you can make out with a hacksaw or hammer and chisel, cutting part way, then breaking. Cutting and placing wire fabric is always an ornery job that calls for gloves.

10

When you cut the fabric to size, allow about 2 inches clearance from the forms all around. Place the fabric so that it is continuous and overlapped—except for a break at any expansion joint that you may be providing. The overlap should be equal to one spacing of the wire, which will be 6 inches in most cases. Thread wire ends together at the laps so they won't try to creep above the surface of the concrete, or tie them together with wire.

Having cut and placed the fabric to make sure it covers the job, you will have to decide how you are going to handle it during the pour. There are several ways. I'll begin with the one that I commonly use because it seems to me the simplest.

With this method you simply leave the mesh lying on the prepared subbase. You pour the concrete over it, and then pull up on the mesh to bring it to the approximate center of the slab. You keep doing this as you go along.

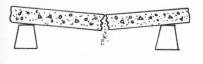

When a slab is supported at two or more points it will tend to break—if it ever does —by pulling apart at the bottom. . .

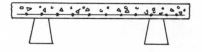

. . . so strengthen it by putting reinforcing steel near the bottom surface where it is needed most.

Concrete supported at one point will be likely to fail by pulling apart and weakening at its upper surface. . .

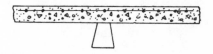

. . . so place reinforcing as high as possible in the slab (but not closer to the surface than 1 inch).

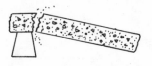

The weakness of concrete in tension is the reason that failure in a cantilever also tends to occur at the top . . .

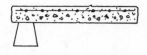

. . . so concentrate the steel reinforcing high in cantilevered slabs (such as elevated hearths) too.

Why reinforce? Here are two driveways built in similar conditions. Wire mesh has helped to hold the concrete together and spread the stresses in one. The other, poured without reinforcing, has been a victim of stresses produced by temperature changes.

If you are using a rake to spread the concrete, you can hook the wire with a tine and pull it up. Or you can file a notch in a spade half an inch from the end, and use this as a hook.

A second method calls for placing bits of old concrete, or small flat stones, under the mesh to hold it up during the pouring.

A third approach is especially suitable for thick slabs and for ones where the positioning is critical. You first pour concrete to half the thickness (more, or less, if you want the mesh at some other point) and then place the mesh on top of this pour. Then you complete the pour on top of it.

Or you can compromise by simply dumping a wheelbarrow load or a few shovelsful of concrete here and there around the area. Let these support the mesh while you complete the pour.

No matter which method you use, there may be a tendency for some of the

Simplest way to place wire fabric is to lay it out in advance and flatten it to the ground. Then, as you pour concrete, hook the mesh and pull it to the center.

wire—and especially the loose ends—to creep above the surface. Watch for this and pound down any escaping wire at once. If you don't they'll be a nuisance during the finishing.

In an ordinary supported slab, the stresses vary. So you'll want the mesh about halfway between top and bottom surfaces. The exact placement is not critical, but strive to keep the mesh at least an inch away from either surface.

When a slab is suspended between points, as it might be with a porch floor, you are concerned with the tendency for it to break by coming apart along the lower surface. In this situation, you'll want a relatively thick slab with the steel quite low in the slab—about an inch from the bottom.

The principle involved is one that becomes very important with concrete that may be under heavy stress. With this you'll probably be using reinforcing rod, so we'll talk about the theory in the next section of this chapter.

When pouring a relatively thick slab, this method of placing steel fabric becomes useful. About half the thickness of concrete is poured first and leveled. Then the steel is rolled out and the rest of the concrete is immediately poured around it.

RECOMMENDED STYLES OF WELDED WIRE FABRIC REINFORCEMENT FOR CONCRETE PROJECTS AROUND THE HOME

Type of Construction	Recommended Style	Remarks
Barbecue Foundation Slabs	6x6-8/8 to 4x4-6/6	Use heavier style fabric for heavy, massive fireplaces or barbecue pits.
Basement Floors	6x6-10/10, 6x6-8/8 or 6x6-6/6	For small areas (15-foot maximum side dimension) use 6x6-10/10. As a rule of thumb, the larger the area or the poorer the sub-soil, the heavier the gauge.
Driveways	6x6-6/6	Continuous reinforcement between 25- to 30-foot contraction joints.
Foundation Slabs .... (Residential only)	6x6-10/10	Use heavier gauge over poorly drained sub-soil, or when maximum dimension is greater than 15 feet.
Garage Floors	6x6-6/6	Position at midpoint of 5- or 6-inch thick slab.
Patios and Terraces	6x6-10/10	Use 6x6-8/8 if sub-soil is poorly drained.
Porch Floors a. 6-inch thick slab up to 6-foot span . b. 6-inch thick slab up to 8-foot span .	6x6-6/6 4x4-4/4	Position 1 inch from bottom form to resist tensile stresses.
Sidewalks	6x6-10/10 6x6-8/8	Use heavier gauge over poorly drained sub-soil. Construct 25- to 30-foot slabs as for driveways.
Steps (Free span)	6x6-6/6	Use heavier style if more than five risers. Position fabric 1 inch from bottom form.
Steps (On ground)	6x6-8/8	Use 6x6-6/6 for unstable sub-soil.

Special conditions, such as warm-air heating pipes to be buried in the slab, call for different procedures. Here you see the mesh placed over the ductwork and cut out around the rectangular connection. As concrete is placed, the mesh is hooked with a notched spade and pulled up to mid-slab position.

REINFORCING ROD. When designing concrete work for strength, there is one very simple principle to keep in mind: concrete is strong in compression but weak in tension. To put it another way, it is very difficult to squeeze cured concrete together, but easy to pull it apart. Concrete makes a good pillar but a poor rope.

Steel, on the other hand, makes an excellent rope but a poor pillar. So steel and concrete work extremely well together—if the steel is placed where its great tensile strength is needed most.

Where should you place the steel? Obviously, at the point where the expected stresses will tend to pull the concrete apart.

Consider a concrete beam or slab that is supported at both ends. What you have to fear is that imposed loads, or its own weight, will tend to make it buckle by pulling apart along the underside. Prevent this by placing your steel close to the bottom.

Now consider a beam or slab that is supported mainly in the middle. Loads will tend to make it break by coming apart along the top surface. So put your hold-together steel near that upper surface.

A cantilever, such as a raised hearth for a fireplace, is equivalent to one half of this center-supported slab. If it breaks it will be by breaking downward, so it too needs its steel placed high.

A slab on the ground, or a driveway or a sidewalk, is supported by the earth or the base material at various places. In effect, it is a combination of

15

both the end-supported and the middle-supported conditions. So it is logical to compromise and place the reinforcing rod or welded-wire fabric at about midpoint within the slab.

With these simple principles in mind, you become a pretty good seat-of-the-pants engineer or ferroconcrete designer.

The principal product you'll be using for reinforcing is what is commonly called re-rod. This comes as 20-foot lengths of steel, usually round but with nubs—called deformations. These help keep it from slipping within the hardened concrete.

For home-sized jobs you will use mostly half-inch-diameter rod, although larger and smaller are readily available. You can cut it by sawing halfway through with a hacksaw or with a flexible cut-off wheel on a circular saw, then breaking it. If you are doing a big job that calls for a lot of cutting, try to have someone do it for you with a torch. Better yet: decide the lengths in advance and have the cutting done with the big cutter to be found almost anywhere that re-rod is sold.

Use full-length pieces as far as possible. Where laps come, the rule is overlap 20 diameters. That means that with half-inch rod the lap should be at least 10 inches. Wire the splices together if possible.

It is usually easiest to place the steel in the concrete during the pour. You may be able to suspend it within high forms.

You can usually bend re-rod around a post or tree, or by putting one foot on it and pulling upward. For heavy steel or short bends, slip a length of pipe over the steel and use this to increase your leverage.

Except in shallow foundations you'll want to run your steel two ways, forming a heavier equivalent of wire fabric. In a cantilevered hearth, for instance, you'll create the heavier equivalent of a 6-by-6-inch mesh, to reinforce in both directions.

With re-rod, as with mesh, try to keep the steel buried so that it is covered by at least an inch of concrete.

CHAPTER FOUR

HOW TO MIX CONCRETE

THE SIMPLEST container for mixing small amounts of concrete is a wheelbarrow. Having mixed the concrete, you can then easily move it to where you want it and dump it directly into the forms.

In the absence of a wheelbarrow, any smooth, clean surface, such as an existing slab of concrete, will do. You can clean it off afterwards with a hose.

You can also use a hand-operated concrete mixer. But a power-operated mixer is the thing for any major job. It will cut your effort in half. The bigger ones usually use a gasoline engine, but for home jobs a mixer with an electric motor is adequate, and simpler and quieter as well. Even the smallest ones will keep two or three people busy enough.

If transit-mix concrete is available where you live and can be had at hours when you want to work, there is not much point to mixing your own for any but the smaller jobs. Why this is true—and how to work with transit-mix—will be covered thoroughly in the next chapter.

MIXER-WHEELBARROW COMBO. For those smaller jobs a good alternative to a regular mixer is the mixer-wheelbarrow combination shown in the photos. Combining the soul of a concrete mixer with the body of a wheelbarrow, it's a homestyle ready-mix truck. Weighing in at 120 pounds with motor, it's far heavier than a 'barrow but lighter and more compact than most mixers. Of course it's much easier to transport and to store than the separate items it replaces.

This mixer-wheelbarrow sells at about the price of a moderate-cost conventional mixer—less than $100 plus freight. It's available through Sears, Roebuck stores and catalog in the West, and from the manufacturer: Allen-Jac Corp., 1190 N. 13th St., San Jose, California. It can be bought with or without a ¼-hp electric motor.

An alternative, for the fellow who only does an occasional mixing job, is to rent a mixer from one of the rent-all centers which have sprung up in most communities.

Since a mixer represents a substantial investment and since its efficiency will make a great difference in your enjoyment of concrete work, you'll want to make your choice with care, when you decide to buy. You may get some help from conclusions reached after a careful use-test of the combination I made for *Popular Science Monthly*.

Setup for mixing concrete: mixer, shovel, supply of sand and gravel, sack of cement, clean water.

If care is taken, measuring by shovelful is accurate enough and easy to do. In building his home, seen here, the author used 80 cubic yards of ready-mixed concrete for slabs, walls, and driveway. An additional 5 yards was put into a big fireplace.

1. Used as a one-man tool, which essentially it is, the barrow-mixer helped me to make and place more concrete in an hour than I have ever done before with any electrically powered mixer.

2. When two of us used it on a fairly large and straightforward job (porch slab), it didn't work us so hard nor get results quite so fast as an ordinary mixer plus a wheelbarrow would have done.

3. On more ticklish jobs, such as casting small blocks and stepping stones, it can keep two men more than busy.

4. Using this combination tool cuts substantially the labor of many small concrete jobs. It saves one whole operation—dumping from mixer to wheelbarrow. It reduces get-ready and clean-up times.

5. Though lighter than you might expect, this mixer is solid and sturdy, perhaps because of its welded all-steel construction. So far as I could judge it should last for a lifetime of casual, around-the-place work.

6. It turns out 1¼ cubic feet of concrete in each batch, slightly less than the capacity of the usual small mixer. This means four or five loads to a bag of cement. But it does a good mixing job in as little as a minute. This helps account for the speed with which it will let you complete a stretch of garden path or other modest job on an idle Saturday afternoon.

Typical small portable mixer like this can be turned by hand or by electric motor. After ingredients are thoroughly mixed, barrel is flipped to dump concrete into wheelbarrow.

7. On delicate jobs where you use up concrete slowly, this combo offers an unexpected convenience. You can keep it churning while you place and tamp part of the batch. That way you don't have to stop and stir the mud with a hoe to keep it plastic as you often must when a partial load sits waiting in a wheelbarrow.

8. Besides being a combination tool, the barrow-mixer is a multipurpose one. You can get extra utility out of it if you have occasion to mix and move liquids or soil and mulch for gardening. I know of one of these devices that is used at a golf course to clean balls. You might find many other uses for this gentle tumbling action, which I checked out at 20 rpm.

Those are the advantages. Now for the drawbacks:

1. Because the machine is tied up during delivery of concrete as well as mixing, it won't keep as big a crew busy as a stationary mixer can. On such simple jobs as porches and other slabs, walls and foundations and sidewalks, it's pretty definitely a one-man machine.

For steps and other fussier tasks, though, it will keep two or even three people occupied—one to mix, wheel, and pour, the others to spread, spade, shape, float, and finish the concrete.

2. Because the mixer must be turning to dump its mud easily, a long extension cord may be needed. It will have to reach to where you are putting the concrete as well as to where you are mixing it. For safety this should be a three-wire grounding cord just as for any portable power tool. The mixer comes with a short cord and three-prong plug.

3. Because it's heavier than a wheelbarrow it's hard to trundle up slopes

When wheelbarrow-mixer combination is used, concrete may usually be dumped directly from mixer into forms, saving one step.

Wheelbarrow-mixer combination is light and compact enough to take from one job to another in the trunk of a sedan or in a station wagon or pickup truck.

or ramps. If you have uphill jobs you'll prefer to use a regular wheelbarrow and treat the combo like a conventional mixer. This is perfectly feasible but it does involve extra trouble.

INGREDIENTS OF CONCRETE. Whatever the mixing method you use, the ingredients of quality concrete are the same: portland cement, fine aggregate, coarse aggregate, and water.

Portland cement is sold in sacks of 94 pounds, and each sack holds 1 cubic foot. Portland cement should be a free-flowing powder. It should not contain lumps that can't be pulverized between your thumb and finger. If you store cement, keep it in a dry place where it will not absorb water and get lumpy.

Fine aggregate such as sand should range in size from ¼ inch down to particles which will pass through a sieve making 100 openings per inch. Do *not* get mortar sand for making concrete since it contains only small-sized grains.

Coarse aggregate is well-graded gravel or crushed rock. These particles range from ¼ inch upward. Coarse aggregate should be sound and hard and not flaky.

Aggregates make up 66 to 78 percent of the volume of finished concrete. It is important that aggregates be clean and not contain lumps of earth or organic matter. Your material supplier will be able to provide good aggregate but you must keep it free of foreign materials on the job. If you use unwashed aggregate from local pits or creeks, be sure that it has been proven to make good concrete.

Mixing water for concrete should be clean and free of acids, alkalis, oils, sulfates, and other materials detrimental to good concrete. A good rule to follow is that water for concrete should be clean enough to drink.

If your freshly mixed concrete looks like this when tested with a trowel, it does not contain enough sand-cement mortar to fill the spaces between particles of coarse aggregate. It will be hard to handle and place and you'll get concrete that may be porous and have rough, honeycombed surfaces.

Trowel your newly mixed concrete and see if it looks like this. If it does, your mixture contains a desirable proportion of cement-sand mortar. All spaces between bits of gravel are filled with mortar, but untroweled parts are still rough. Concrete like this is economical and easy to work with, and it will give you first-class results.

Here is how concrete looks when it contains more than the necessary proportion of cement-sand mortar. This concrete is plastic and easy to work and will produce smooth surfaces, although it may be somewhat porous. Principal objection to it is that it uses too much cement and so costs you a little more than necessary.

The amount of mixing water used per sack of cement determines the durability, watertightness, and strength of concrete. Generally, less water used per sack of cement means better quality concrete. Some concrete jobs must be stronger and more watertight than others. So the intended use of the concrete should determine the amount of water to use per sack of cement.

In making good concrete, it is important that the fine and coarse aggregates be proportioned so that the finer particles will fill the spaces between the larger ones. Your material supplier may have aggregate that has fine and coarse material mixed in the proper proportion for concrete.

In general, the most economical mix is obtained by using the largest size coarse aggregate that is practicable. This is usually about one-fourth to one-fifth of the thickness of the concrete in which it is used. For example, in a 4-inch slab or wall you would use coarse aggregate up to about 1 inch in size.

Technical manuals on concrete always specify the amount of water to use. This depends upon what the concrete is being used for and also upon how wet the sand, or sand-and-gravel mixture, is to begin with.

For noncritical work such as footings and foundation walls and retaining walls, you should use 6¼ U.S. gallons of water for each sack of cement if your sand is merely damp. If the sand is wet, decrease the water to 5½ gallons. If it is very wet, add only 4¾ gallons of water.

For more important work, such as basement walls, slabs, sidewalks, stepping stones, and curbs, you should reduce each of the figures above by one-half gallon.

For heavy footings and walls where neither strength nor waterproofness need be maximum, standard concrete consists of 1 part of cement to 3 parts sand and 4 parts gravel.

A richer concrete—using proportionately more cement—is desirable for most other work. For sidewalks, driveways, steps, floors, and the like, use 1 part portland cement to 2 or 2¼ parts sand and 3 parts gravel.

You can measure the ingredients most easily by using an ordinary water bucket. Or you can make a wooden measuring box with inside dimensions of 12 by 12 by 12 inches. Filled to the top, it will give you an accurate cubic foot.

This is the precise and recommended procedure, but I will confess that I never do it this way, and I have never known an amateur concrete worker of experience who did. Or a professional either, on the rare occasions when one is not using transit-mixed concrete.

If I am doing a little job without a mixer, I proceed this way.

I put two shovelfuls of sand into a wheelbarrow or onto a concrete slab. I add one shovelful of cement. I mix these thoroughly with the shovel or with a hoe. Then I add three shovelfuls of gravel and mix again. Then I make a depression in the center and add water from a garden hose, a little at a time, mixing as I go. I continue until all ingredients are well combined and of the desired stiffness. If I find I have carelessly put in too much water, I mix in a little more sand, gravel, and cement.

I follow the same steps when using a power mixer.

If the aggregate has come already combined, as it usually does in my part

Is there too much water in your concrete? If a cupful of it turned over holds its shape almost perfectly, as does the specimen at the left, it is just right for most purposes. Concrete that slumps partially, as at center, will be more easily workable for a floor slab or a thin wall. A mix that collapses into a puddle, as at right, is too sloppy to produce strong, watertight concrete. Use less water in your mix.

of the country, I simply mix this mixture with portland cement as the first step.

Of course this method is not very precise, even when you try to keep your shovelfuls uniform. So stay on the safe side by keeping your mix comparatively rich. For each shovelful of cement use two of sand and three of gravel (or five of combined aggregate).

It won't take you long to learn to judge the consistency of concrete by looking at it. The photographs offer a good guide to learn from.

HOW MUCH TO BUY. In buying the ingredients from which to make concrete you'll probably find two sources of supply: sand-and-gravel companies and retail dealers in building supplies. The gravel company will usually be cheaper for fairly large amounts of aggregate, say a couple of cubic yards or more. A lumberyard or landscaping-supply place may be more economical when you need only a cubic yard or less, especially if you can haul it yourself. Protect the trunk of your car with corrugated cardboard or a few layers of plastic sheeting that you can reuse for curing concrete.

To estimate how much sand, gravel, and cement you'll need for a job, first figure the volume of concrete. As described in more detail in the first chapter, you just measure the slab or wall in feet and multiply the three dimensions together to get the volume in cubic feet.

Then buy sand and gravel amounting to at least 20 percent more than that number of cubic feet. This is to allow for waste, shrinkage, and imperfect measuring. If you find, for instance, that you need 22 cubic feet of concrete, buy a full cubic yard of aggregate (sand plus gravel), even though this is 27 cubic feet. If there is a little left over, it won't have cost you much, if anything, extra. Running short is a terrible nuisance.

As for cement, a standard bag holds 1 cubic foot. If you're making a good, rich, general-purpose concrete (1 part cement to 5 aggregate) you will then need one sack of cement for each 5 cubic feet of sand and gravel. Or five sacks of cement for each yard of concrete wanted.

CHAPTER FIVE

HOW TO WORK
WITH TRANSIT-MIX

IF YOU'VE never used transit-mixed concrete—the stuff that comes in the big revolving-barrel trucks—the idea may make you a little nervous. You may, quite reasonably, fear that the stuff will be hard to handle. Or more expensive than the kind you mix yourself. Or that the big ready-mix companies aren't interested in a customer with only one sidewalk or patio to pour.

Actually, none of these things is true of most jobs calling for a cubic yard or more of concrete. If you take the precautions and make the preparations to which this chapter is devoted, you shouldn't have any trouble. And your effort will be far less than when you mix your own because you will have eliminated half of the job.

The cost advantage or otherwise will necessarily depend on local conditions. You can check it as I did—or you can assume the comparison will work out much as it did in my locality.

Begin with the price per yard of sand-and-gravel mix, delivered to your site. Add about 50 percent. You must do this because, oddly enough, it takes a yard and one-third of aggregate to make a yard of concrete, and you'll waste some if you have it dumped onto the ground.

Now add on the delivered price of five sacks of cement. Compare your total with what your nearest transit-mix company charges for a yard of five-sack concrete.

When I last checked figures in my area, I found that the ingredients of a yard of concrete added up to $23.90. The quotation on ready-mixed concrete was $21. It seems that if I wish to mix my own I must expect to pay about $3 for the privilege of shoveling two tons of sand, gravel, and cement.

Most driveway, patio, or sidewalk jobs around the home use up a lot more than a single yard of concrete. Quantity price, which often means from three yards up, is considerably less—currently about $15 a yard where I live.

There are a few principles by which you should be guided in using transit-mixed concrete. Follow them and you're not likely to run into serious difficulties.

1. Get the soil ready ahead of time. Level the area by shaving off the high spots and filling in the low, then tamping to compact the soil. If you have

26

clay, hardpan, or poor drainage, fill with a few inches of gravel or cinders.

Chop off any nearby high spots outside the forms too. You don't want hummocks in your way when the concrete is pouring in fast and you are running a 2-by-4 across to strike it off level.

2. Build your forms level and stout. You'll be too busy at pouring time to adjust their level. They must be extra sturdy because you'll be dumping several tons of mud within a few minutes. And if they collapse before the truck driver's eye your embarrassment will hurt you worse than the loss of concrete.

Brace wall forms from the outside so they won't spread and tie them together, too. If they show any tendency to collapse inward, stick in short pieces of 1-by-2 as spacers. You can pull these out as the concrete goes in.

In the case of a slab or sidewalk, pound in or saw off all stakes level with the top of the form. Then they won't impede you when you saw off the concrete to a level.

Remember that a floor slab much more than 10 feet wide is hard to strike off. Divide big slabs in the middle with a 2-by-4 or length of steel pipe at the same level as the outside forms. As soon as you've struck off the slab to level, pull out the divider and fill the hollow with concrete.

If you are going to use the same form material over again, brush it with old crankcase oil so concrete won't stick to it. If you plan to salvage the form material for another use where oil will be objectionable, shellac it instead.

But it isn't absolutely necessary to treat form lumber or plywood if you can pull it off within about a day and scrub it then.

3. Cut and shape your reinforcing. Middle of a pour is no time for hacksaw or nippers. Place as much of it as you can in advance. Rod can often be supported inside forms on long nails. Mesh can sit on little chairs, either the kind made and sold for the purpose or ones you can dig up. Small chunks of brick or concrete or even stones will hold the mesh up to the height of the middle of a slab.

Instead of using chairs, stones, or other support, you can simply put the mesh into the area and lift it into center-slab position with a rake while the concrete is being poured. But do it as you go along; don't let it go till there's too much too-stiff concrete sitting atop it.

If you plan to pour quite a lot of mesh-reinforced slab you'll find it worthwhile to prepare a flat-bladed spade to do this work. You can use the same spade for general compacting of the concrete. Simply saw or file a notch in the side of the blade half an inch from its end. You will be able to hook this over the wire and lift it quickly.

If you're pouring a slab or foundation that must hold sill bolts, be sure you have them ready.

4. Prepare your truck route. Scheme things to bring the truck as close to the job as possible, then go over the route to look for weak spots. You may want to plank some of your driveway or lawn with 2-by-12's. Consult the transit-mix supplier if you see any problems. The company may send a man to look things over, and—in my experience—he may even lend you some timbers if he finds you need them.

If the truck-route problem is traction, such as soft sand or mud, the mix people will want to know about that too. They may, as they did for my last job, be able to bring a stack of airplane landing mats to spread out under the wheels.

As far as possible, arrange to dump the concrete right from the truck into the forms. Sometimes you can do this in difficult spots by preparing a wooden chute or one made from a sheet of corrugated metal. If you'll have to wheel concrete, make a sturdy plank runway ahead of time.

You can save a lot of sweat on any long slab or driveway by planning for the truck to drive right into the area. Make one end form removable to let the truck in, but have it all ready to put into place.

Keep in mind that a loaded mixer carrying 5 yards of concrete weighs about 25 tons. A long wheelbase and dual wheels spread the weight well, but you can see why the ready-mix companies assume no responsibility for damage done once they have passed the curb line onto your property.

In allowing clearance, figure the width of a transit-mix truck as 8 feet and its height as 12 feet.

5. Assemble tools before you start. These will be the usual needed for any cement job—shovel and rake for placing concrete, board for striking it off level, wood float for leveling it further. You also will likely need a steel trowel for finishing, possibly a broom for texturing the surface, and an edger for rounding off corners. To divide or pattern a sidewalk you may want a groover.

Consider also the virtue of having boots or overshoes ready in case some of the spreading can best be done by wading right into the mess. Personally, I wear overshoes over bedroom slippers, which is very comforting to bunions. Gloves are good to have at hand too, and preferably spares so you can change if they get wet.

And you will almost certainly need a wheelbarrow. If it is a big job, a contractor's deep model will beat the ordinary house-and-garden type all hollow. But don't rush out to buy, or even rent, one. See first if your ready-mix man will, like mine, lend you one free of charge, hauling it along when he brings the mud. He probably won't, though, unless you arrange for it in advance. It's one of those valuable free services you can latch onto only if you know about it. Likely nobody will tell you.

The same goes for an extra-long chute. If you discover in advance that you need one and remember to ask for it, your driver can bring it along. Having it can sometimes save you an hour of backbreaking shoveling.

6. Make a runway for that wheelbarrow. Use as broad and stout a plank as you can get, and slope it as gently as possible. Using ready-mix is a prime sweat-saver, but not if you have to hustle big barrow-loads up a steep and narrow.

In laying out the runway, fix it so you'll pour the first concrete at the far end of the day's job. You don't want to have to go round or wade through the first load with every one thereafter.

Make your barrow highway as smooth as possible too. This is not so much to save jar on your tennis elbow as to keep the concrete from separating. You wouldn't think such heavy stuff could be so delicate, but it is.

Using transit-mix concrete saved the author hundreds of hours of hard work in building this slab-floored, concrete-and-stone-walled house overlooking the Pacific Ocean.

For the same reason, schedule your wheeling to dump the stuff where you want it without too much shoveling around. You'll want to be able to place each load so it will slump into the previous one, not away from it or on top of it.

Plan things so that when pouring a wall or deep foundation you can lay the concrete in layers six inches to a foot thick, starting at the ends and working toward the middle.

7. Ask for help in advance. Most concrete jobs call for at least two people—one on each end of the screed board. Make the deal with your neighbor or hired helper so you can be sure he'll be ready when the mud is. There's nothing much readier than ready-mix.

Main thing is to have enough help to avoid the extra charge companies commonly make for stand-by time if you keep the truck waiting around too long. I've rarely been charged this, even on the sad occasion when I ignored Point 2 and kept a driver working an extra couple of hours. But your company may not be so generous.

A few sturdy bracing posts, nails, and a hammer should be right at hand,

29

too. If a form starts to give, there may not be time to call for a helper nor a chance to go searching for bracing materials.

If there's no neighbor to lend a hand, a reasonably sturdy wife will serve. With the devoted help of mine, I was able to pour a concrete wall 85 feet long and 5 feet high in a single day recently . . . some 15 yards of concrete plus a couple of tons of stones we set into it as we went along.

The secret in this was the help we got from the third member of the team— the fellow who delivered the ready-mix. He worked steadily the whole time, shoveling the mud from wheelbarrow into forms whenever it wasn't feasible to pour directly.

Maybe some drivers won't pitch in and labor by your side, but all four that have served me in recent months did. And so did the salesman from their company when he happened to drop in to check up at just the moment when another pair of biceps was needed. So there's another free service you may get if you need it.

If you're lucky enough to have a couple of teenagers in the family, they too can be valuable members of the crew at difficult or rushed times. Maybe pouring concrete is more interesting than household chores, but mine always made themselves useful as soon as the truck arrived.

8. Estimate your needs carefully and place your order early. If you place your order a day or two ahead you'll probably be able to pick your time. First load of the day, around 8 o'clock, is usually best. Then you'll run no risk of troweling by moonlight, which is a heck of a way to spend an evening and also leads to rough results, since you're pretty tired by that time.

Estimating is just a matter of measuring, and using a little grade-school arithmetic. Multiply the length and width in feet by the thickness in inches and divide by 12 to get the cubic footage. You'll need a yard of concrete for each 27 cubic feet.

If the figures throw you, or you have some other problem, such as whether the truck can get into where you want it, pass word along to the cement company. Very likely they have a man who looks into just such matters and will be glad to come in advance to advise you.

It may pay you to figure your job so you use enough concrete to get the best price. In my area, 2 yards bought separately cost a trifle more than 3 yards taken at a crack, which works out to an appealing 33 percent discount on the latter.

If you are doing a lot of concreting—building a slab-floor house, say, or a long driveway—investigate these price brackets fully. My dealer, for instance, gives me the quantity price even on days when I order an undersized batch, because he figures it on the basis of my consumption over the weeks.

Also consult the company on what kind of mix to use. You'll want what is most economical but strong enough to do the job and meet any local building-code requirements. One advantage of using ready-mix is that you can usually get expert help in engineering.

For example, a common kind of ready-mixed concrete is what is called five-sack. This means that the equivalent of five standard sacks of cement have been used in making each cubic yard. But if you are pouring a heavy foundation or a wall that is far thicker than it needs to be for strength, a leaner-mix—one that contains less cement—will be good enough. The dealer may then suggest that he bring you a four-sack concrete, at a saving to you of perhaps a dollar a yard.

If you are pouring into forms of a difficult shape, you can get especially plastic mixes. If there is a danger of your concrete setting up before you'll have time to spread and trowel it, you can use a slow-setting concrete. There are additives designed to protect you from too-fast setting, likely in dry areas or on particular hot or windy days.

Or you may have the opposite problem—a likelihood that the concrete will set up so slowly that you'll be troweling by electric light at midnight. Quick-setting concrete will protect you from this distasteful prospect.

Scheduling your concrete work for moderate days is the best answer, however, to most of these problems. Just as the rule is that concrete should be mixed with water fit for drinking, so should you try to do the work on days for human beings. Very hot days are not the best, and ones that threaten a freeze are worse. Concrete likes days hovering around 70 degrees just as well as you do.

A final tip on ordering: order enough. Running even a shovelful short is an infernal nuisance on most jobs. You will, however, be able to make up for a slight shortage on most kinds of pour by having a few rocks on hand to shove in.

9. Prepare some extra forms in which you can use up any excess concrete. Then you won't lose anything by ordering generously.

Most yards need various little concrete items—splashboards, garbage-can bases, bits of sidewalk. Have forms ready for them. Or put in preliminary forms for any future jobs.

One simple solution is to build a few rough forms from 2-by-2 lumber, perhaps 2 feet square. Fill them with excess concrete and you'll accumulate an invaluable supply of stepping stones ready to deposit wherever you need them. If you hose down the surface of these stepping stones just as the concrete begins to set, perhaps an hour after pouring, you will get a particularly attractive textured surface of exposed aggregate.

10. Wet the area thoroughly. Do this a couple of times on the day before you pour concrete. Do this again on The Morning unless it is still noticeably damp. This will keep the earth from sucking all the water out of the concrete, something that leads to what concrete workers commonly refer to as a blankety-blank snap set.

What this amounts to is that the concrete dries so fast you don't have time to work it into position easily and to finish it before it has taken on a shape of its own sweet choosing. Losing moisture into the ground also prevents the slow curing that produces strong concrete.

If you find the mix a little hard to handle, the driver will willingly add as much water (he has a tank of it on his truck) as you ask for. Don't be tempted a bit farther than is absolutely necessary in this direction. A soupy mix is easy to handle, but it produces weak concrete.

Ready-mix drivers are, in my experience with a couple of dozen different ones, amazingly helpful skillful people. I've yet to meet one who wouldn't grab a shovel or the other end of a screed board and sweat along with you when needed. They'll warn you if your forms aren't sturdy enough and drive a quick nail to strengthen a form that gives a warning squeak. There's no law that you can't reward and encourage a driver with something tall and icy (but nonalcoholic) if the day is hot.

Make the right preparations, cooperate with your ready-mix people—and you'll find that half the backache and all the headache has gone out of your concrete jobs.

CHAPTER SIX

HOW TO USE
YOU-HAUL CONCRETE

FOR THE home concrete workers, haul-it-yourself concrete has proved to be a most welcome innovation. It's the greatest thing, one tired shoveler remarked, since sliced bread. If it is available by now in your area, you'll find it the ideal solution to getting concrete for middle-sized jobs.

For small jobs, dry-mix concrete in sacks is the simplest thing to use. For big ones, there's transit-mix. But when you need between a quarter of a yard and a yard of the stuff, this new kind of ready-mix that you haul yourself is ideal.

Beginning in the mid-1960s on the West Coast, you-haul concrete has been offered in three versions: trailer haul, trailer-mixer, and pick-up mixer.

If what your locality offers is the first kind, you will drive up with your own car, have a specially designed box trailer hitched to its bumper while a small batch plant begins to mix your mud. The stuff that will then be poured into the trailer for you to haul away will be concrete with a couple of extra ingredients. These will be chemicals designed to slow the setting of concrete— to give you time to haul it home and place it—and to keep the gravel from settling to the bottom as you jounce it along the highway. Concrete being heavy stuff, the trailer will be equipped with a hydraulic hoist for easy pouring, and surge brakes that apply themselves whenever your car slows up. There may be a few things clinging to your trailer. You-haul outlets offer for rental at prices from 25 cents to about a dollar such items as shovels, wheelbarrows, tamping devices, and trowels.

On the other hand, what you may be hauling behind your car is a "cannon." That's the appropriate name given to the second of three kinds of devices in use. It's a trailer-mixer and it does rather resemble a cannon. With the cannon there is no problem of aggregate settling out or concrete setting up too fast on the way. The dealer needs only hoppers for handling materials instead of the costly (up towards $20,000) batch plant. And he doesn't run up against the zoning regulations that prohibit batch plants in many areas that permit ordinary building-materials dealers.

33

Add a trailer-mixer "cannon" like this to your car and you have your own transit-mix truck.

Both the cannon and the you-haul-it trailers share certain drawbacks. Either one means a heavy load hitched to your bumper, although the tandem wheels of the cannon do take most of the weight off your car. And you'll still need some skill in backing a trailer, not the easiest job in some driveways and backyards.

You can avoid these problems if you can find a dealer offering the third form of you-haul concrete. This is a mixer-cannon—mounted on a heavy-duty pickup truck. Striving for the ideal answer, a do-it-yourself lumber dealer in Santa Cruz, California, acquired a U-Pour cannon without understructure early this year. Mounted on a second-hand heavy-duty Ford pickup, it is now offered to customers who want to haul their own ready-mix without pulling a trailer.

The price you can expect to pay for your concrete will be about the same, whichever of the three systems you find in your neighborhood. Typical charge is $20 for a one-yard load, possibly a couple of dollars more on weekends—when some dealers find they have eager buyers waiting in line. Figure on about $12 for a half yard, $8 for a quarter. All prices include trailer or truck rental, though you may be nicked an additional half a dollar for use of a hitch if your car doesn't already have one.

Handling the concrete will give you no problems if you've had experience with transit-mix—and don't make the grievous mistake of disconnecting that heavy trailer from your car while it is charged with the mud.

Main precaution to take is simply to have everything ready before you go for the load. Make sure tools are ready and forms are stoutly braced so you can pour the concrete rapidly and get the equipment back to the dealer without an overtime charge. You may also be penalized if you neglect to hose the trailer or mixer clean.

CHAPTER SEVEN

HOW TO PLACE
AND FINISH CONCRETE

THIS CHAPTER is about one of the most common kinds of concrete work around the home: placing concrete flat on the ground in forms and giving it a smooth of semismooth finish.

What you get this way is utilitarian concrete. It is durable. It is easy to sweep or hose off. It makes a good garage or shop floor or a smooth base for covering with resilient flooring such as vinyl-asbestos tile.

For most other purposes—including many for which it is still being used far too often—it is not the most attractive choice. It lacks color and texture. I strongly recommend that you consider some of the alternative treatments des-

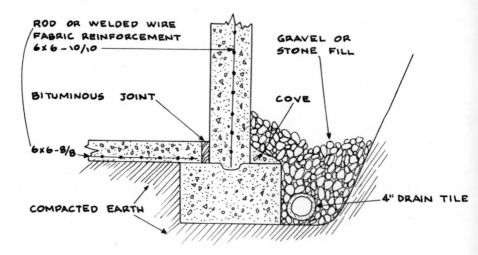

A typical basement slab for normal soil conditions is made like this. First cast the footing with a groove in the top, formed by a beveled 2-by-4. Pour wall and slab on top of this as shown, with a space between for waterproofing. Cove, gravel, and drain tile on the outside will usually prevent water from seeping in. A slab floor for a house or cabin may be cast in much the same fashion, although you will not usually need the outside drainage arrangement shown here.

Tamper is used to compact the subgrade so the concrete will have uniform support. Wooden fence post or a short length of 4-by-4 lumber can be used instead.

Jitterbug is used to force large gravel slightly below the surface for easier finishing later. You'll need it only with very stiff mix — and usually not at all.

cribed in other chapters. But you will still follow the basic methods described and shown here.

Begin by removing all sod and vegetable matter from the area. Dig out any soft places and fill them with sand, gravel, or crushed stone, thoroughly tamped. Fill low areas and tamp the fill. Local conditions may require using 4 to 6 inches of well-compacted sand, gravel, or cinders over the whole area.

You will usually want to place form lumber at the same time that you level the subgrade. Forms for most on-ground concrete work are simply 2-by-4's on edge, held in place by stakes of 1-by-4 or 2-by-4 lumber.

In using a level to establish position of forms, keep in mind that outdoor concrete should seldom be dead level. A driveway or patio, and often a porch, should pitch slightly so that rain and water from hosing off will not accumulate. For better drainage, a driveway or sidewalk should usually have its top surface about 2 inches above surrounding areas.

Before you begin placing concrete, dampen the area with a hose. This will produce slower drying and better curing. This is unnecessary, of course, if you are putting a layer of polyethylene sheeting or other moisture barrier on the ground to insure a dry slab.

Edger rounds the corners of a sidewalk or driveway slab. Use it after concrete surface has been struck off level, running it back and forth between concrete and form.

Groover is used just before floating, and often again right afterwards, to produce joints across sidewalk or other slab. Run it back and forth against a straight board.

Wood float is used in finishing all slabs. With it you level the concrete and give it a rough finish, either final or preliminary to steel troweling.

Steel trowel is swung in sweeping arcs to give a slab a very smooth finish. Hold it flat and or use it with a swirling motion for a smooth matte finish.

You can test the quality of your mix with one sweep of a steel trowel. It should be stiff enough not to collapse into a puddle, yet plastic enough to smooth out like this. Homemade wood float is seen in background. These are typical sidewalk forms with reinforcing wire mesh placed and ready to be pulled up to center of slab during pour.

Place your reinforcing wire mesh as described and shown in an earlier chapter. Remember to hook it up into the center of the slab as you pour if you are using this method.

Whether your concrete arrives by transit-mix truck or by wheelbarrow with you behind it, try to place it as near to final position as you can. Excessive raking and shoveling will bring excess water and fine materials to the surface, and this may later lead to scaling or dusting. A rake is the most useful tool for moving concrete about and bringing it approximately to the level of the forms.

As soon as possible, begin striking off the concrete to level. Do this by resting a timber, such as 2-by-4, across the forms and moving it back and forth as you advance. One man can do this reasonably well on a sidewalk job, but

Swing a darby like this to level your slab after striking it off with a seesaw board. This tool is an alternative to a bull float, either being used to make the hand floating easier.

Wood float is swung back and forth across the surface of the concrete after it has been struck off level. Except for edging, this is only finishing usual with sidewalks.

for a driveway or a floor slab it takes two. If the first pass of the straightedge, or strikeoff rod, produces noticeable waves or humps, go back and start again.

You may now go over the surface with a hand tamper, or jitterbug, to force coarse aggregate slightly below the surface. This step is usually not necessary, and I recommend you omit it unless you are using very stiff concrete and have been having trouble finishing it satisfactorily.

Next you should use either a darby or a bull float—if you have either of these tools. A bull float—just a short board with a long handle—is generally used outdoors; a darby is handier where space is constricted. Either is used to bring the surface closer to dead level quickly. If you omit this step you will have to do more work with a wood float and probably won't get bumps and hollows quite worked out.

Long-handled float is used for preliminary smoothing of a slab too large to reach with a hand float. You can make one quickly and easily by attaching a long handle to an 18-inch piece of 1-by-6 lumber.

The slab you see here has been leveled and made semismooth with a wood float. Final finishing is done with a steel trowel. Elevation of leading edge is exaggerated here.

You will probably want to round off the edges by running an edging tool back and forth. If you are grooving the slab at intervals with a jointing tool, do this now.

Next use a wood float to even the entire surface. The uniform, slightly rough texture you will produce is suitable for a sidewalk or driveway, and you can stop at this point.

Steel-troweling is the next, and final, step in finishing. It is a fairly tedious operation—and it comes at just about the time when you'd rather be bathed and sitting around in robe and slippers—so get along without it if you can. It is a necessary operation, however, when a really smooth surface is needed—as for a garage floor or for later installation of resilient tile on a house slab.

Where most people go wrong in troweling is in starting too soon. You can't do a good troweling job until the concrete has begun to stiffen and all water sheen has left the surface. This condition may be reached almost as soon as you have completed the floating—or it may not occur for several hours.

Such a delay will be produced by overly wet concrete, wet subgrade, cool and damp weather, absence of sunshine, or the presence of a moisture barrier, such as polyethylene sheeting, under the slab.

For the first troweling, keep the blade almost flat against the surface. An old trowel is better than a new one; its rounded edges will be less likely to dig in. Trowel out the marks left by the edging tool.

For a slab that is to be broomed or otherwise textured, as described in the

41

A pair of professionals demonstrate position and equipment for troweling a slab to a smooth finish. First troweling is often done immediately after floating with wood floats these men are using to lean on.

next chapter, one troweling is usually enough. But to get a very smooth surface you will probably have to trowel two or more times. Swing the trowel in long, sweeping arcs with the leading edge raised just slightly. Use kneeling boards, as shown in the photo, to reach parts of the slab far from the edges.

All concrete must be cured if it is to be strong and hard. This means that its moisture must be kept in for at least three days.

How you do it doesn't matter. In very cool and damp weather you may find it is possible to keep the surface continuously wet by frequent sprinkling. But it is usually easier and far safer to cover the fresh concrete with inexpensive polyethylene sheeting, weighting edges with earth or lumber or bricks to keep it from blowing off.

Or you can shovel on some clean sand and dampen this occasionally. Even damp earth will do the job—if the slab is later to be covered with some kind of flooring so that earth stains are of no consequence.

Don't do concrete work when a freeze is to be expected. If one comes suddenly after concrete is placed, protect it as well as you can. A thick layer of straw is good insulation. Dirt, preferably over polyethylene film, will help too. Very hot weather is also to be avoided.

CHAPTER EIGHT

HOW TO TEXTURE CONCRETE

THERE'S NO need to be satisfied with plain smooth concrete in places where pattern or texture would look better. In my opinion, this is everywhere. Ordinary floated or smooth-troweled concrete is best reserved for areas that must be easily cleaned, such as garage floors, or for slabs that are to be covered with other flooring materials later. It is actually easier to texture a slab than to trowel it to perfect smoothness. Fine troweling is hard work.

Swirling is the method for introducing a moderate amount of irregular pattern into a concrete surface. For the least degree of pattern, do this when the slab is ready for final troweling. Hold the steel trowel flat on the surface and finish with swirling motions.

If you want a more visible swirl, attack the surface while it is still a little too soft for final troweling. Use an aluminum or magnesium float for a medium texture, or an ordinary wood float for a coarser pattern.

With a brush or a broom you can produce a different kind of texture. The coarser the tool and the softer the concrete when you use it, the more pronounced the effect will be.

If you use a soft push-broom or large paint brush on a full troweled slab you will produce a pattern so slight that it is merely a softening of the slick surface effect.

But a kitchen broom dragged across a slab that has had only an early preliminary troweling will give a very rough surface. In addition to being attractive to the eye, this rough surface is desirable for creating nonslip sidewalks in damp areas and for preventing skidding on driveways with considerable slope. Brooming or brushing can be done in straight lines or, for somewhat different effects, in curves.

You can even produce a travertine finish, a remarkable stonelike effect, without too much effort. After you have poured your concrete and struck it off level, go over it with a darby or a wood float. Then roughen the surface with a stiff broom. This will prepare it to bond well with the finish coat.

Make this by mixing one part of white portland cement to two parts of sand and a small amount of color pigment. Yellow is the usual color, and you will need about a quarter pound of mineral oxide pigment for each bag of cement you use. Add enough water to produce a soupy mixture like thick paint.

43

For a lightly swirled pattern do your final troweling like this. Hold the steel trowel flat against the surface of the slab and move it around in swirling motions.

When swirling is done with a wood float while concrete is still fairly wet, a coarser pattern is produced. Edging and jointing of this sidewalk was done before floating, then repeated afterwards.

Dragging a soft brush straight across a moderately wet surface produces a soft pattern of parallel lines. For even less texture wait until the concrete has hardened more and trowel it smooth before brushing.

It is also easy to create a wavy pattern when brushing the surface of soft concrete. Even a garage floor is improved by light brushing to make it both more attractive and less slippery than a smooth finish.

A coarsely broomed concrete surface is less dazzling than a troweled one in sunshine, far safer when wet. When a patio is cast in a redwood grid, alternate squares should be broomed in different directions for a subtle checkerboard effect. Note that brooming makes it unnecessary to do a thorough troweling job, a tremendous saving in effort.

Travertine effect is produced by dashing on colored mortar and troweling it in, as described in the accompanying text.

Scoring adds an interesting pattern to what otherwise would be an all-too-con-spicuous expanse of plain concrete driveway.

Dash quantities of this colored mortar onto the slab surface with a large brush. You should aim to produce an uneven surface with ridges up to about half an inch high.

Let the surface harden until you can safely put a kneeling board on it. Then trowel the slab to flatten the ridges, leaving the surface smooth in some places and rough in the low spots. The final effect can vary a good deal, depending upon how much mortar you have thrown on and how much troweling you do.

For an even more surprisingly unconcretelike effect, score the surface into random geometric shapes before curing, either as seen here or as shown in the picture sequence on scoring a sidewalk.

The scoring process, which can be done with ordinary troweled concrete as well, is shown in photographs and described in the captions that accompany them. Akin to it is patterning with leaves or circles, also shown in the photographs.

Another ingenious trick for creating texture uses, of all things, rock salt. You simply scatter the stuff over the troweled surface and then press it in with your trowel. After the concrete has hardened you wash away the salt with water, leaving pits or holes in the surface.

Neither this nor the travertine technique should be used in areas subject to freezing water. Water can be trapped in the recesses of these finishes, and when it freezes will cause the surface to break up.

Flagstone pattern is produced by tooling concrete after smoothing it with darby or float. Use an 18-inch length of ½- or ¾-inch copper pipe bent into a slight S shape.

Tooling must be done while concrete is still plastic so that coarse aggregate can be pushed down. After water sheen disappears from surface, float and trowel the slab. Run jointing tool again and then follow up with a final troweling.

Complete the job by lightly brushing the entire surface and carefully touching up the joints with a soft-bristle paint brush, used dry. Then cure as usual.

A simple tool for adding pattern to a slab immediately after final troweling is a coffee can. Cans of several sizes can be used to produce a random pattern like this one.

A leaf pattern giving the effect of fossils is produced by pressing leaves into the surface immediately after troweling. They should be completely embedded but not covered, then left till concrete is stiff. Leaf pattern is most effective if used as a border or as an occasional group or spray to form a focus of interest.

Exposed-aggregate concrete is produced by hosing off the surface after the concrete has begun to set. Ordinary sharp aggregate gives the effect seen here. Improvement produced by adding a few flat pebbles to the surface before floating can be seen in pictures in the chapter on building a grid patio.

Of all the methods of introducing texture, exposing the aggregate is the simplest. Essentially it consists of letting the lightly troweled surface of the slab set up partially and then hosing it off to expose some of the rocks it contains. Here, in more detail, are the steps.

1. If possible, use aggregate containing rounded rocks rather than sharp ones. Transit-mixed concrete companies sometimes offer this alternative if you ask for it, often at no extra charge.

2. Mix and place the concrete in the usual way. Strike it off level and float it as usual. Trowel it while this is still fairly easy to do. You need not do a very thorough job.

3. When the surface has begun to set up, hose it off with a fine, gentle spray until rocks show up. If necessary, scrub with a broom.

4. Come back a little later and hose off the slab again to wash away water containing cement. Then cure as usual.

You can enhance the appearance of an exposed-aggregate slab by adding a few flat pebbles as you strike off the surface. The more pebbles you add the closer it comes to being pebble concrete. This is covered in the next chapter.

CHAPTER NINE

HOW TO MAKE
PEBBLE CONCRETE

IF THERE is one thing above all that I hope this book will accomplish, it is to encourage the use of pebble concrete in place of the ordinary kind. You can get somewhat the effect of pebble concrete by using the exposed-aggregate technique described in the previous chapter. This is true especially if you can get aggregate consisting mostly of rounded or flat stones rather than sharp crushed gravel. However, pebble concrete is still better. It gives you a choice of color and shape and also permits the use of handsome stones that would be too costly to use all through a driveway, sidewalk, or patio slab.

Dealers in transit-mixed concrete commonly sell bags of pebbles for use in pebble concrete. So do many dealers of building and patio supplies. For prices varying from $3 to $10, you can buy a 100-pound sack of large or small flat pebbles selected for color or combinations of colors—black, white, red, and so on. And a sackful goes a long way.

To get more mileage out of expensive pebbles, you can work a combination of pebble and exposed-aggregate concrete. Just add a comparatively few pebbles to the surface. Then when you hose it down to expose them you'll also expose some of the aggregate. There are pictures of this in the chapter on making grid patios.

Where to use pebble concrete?

It is ideal for any driveway—and the longer and wider the driveway the more you need the pebble effect to keep it from looking like a city street.

It is an excellent choice for a sidewalk. In my town of Carmel, California, even the sidewalks in the business district are now being made of this stuff, and a great improvement it is too.

Use pebble concrete for patios and porches too, unless you are concerned about its roughness for furniture placement.

Many of the expensive new houses in my area are using it for entryway floors inside the house.

I even used it for two fireplace hearths in my home. You'll see one of these among the fireplace pictures near the end of this book.

The pictures and captions of the following pages will show you how easy it is to do your own home concrete work in this versatile medium.

To appreciate the value of pebble concrete, visualize this expansive driveway done in ordinary concrete. The dark pebbles used reduce glare and conceal staining both by adding texture and color.

Where you want variety, don't hesitate to mix large pebbles with small ones, pebble concrete with smoother surfaces.

Existing areas of smooth concrete in your yard need not stop you from doing additional sections with a pebble finish.

Here a large patio of pebble concrete surrounds a formal outdoor dining area delineated by use of white concrete.

1. Immediately after placing your concrete and striking it off level with a straightedge board, distribute pebbles over the surface. You can cover the entire surface, as shown here, or use comparatively few pebbles for quite a different effect.

2. Press the pebbles firmly into the surface of the fresh concrete. You can do this with a darby or with an ordinary plank—such as the 2-by-4 you may have used in leveling the slab. If concrete has begun to set, you may have to apply pressure or pound pebbles in.

3. If the concrete has become somewhat stiff, you may have to use a wood float, and bear down hard, to force pebbles below the surface. Continue till surface is smooth and almost free of holes. Foreground area here still needs working over with float.

4. When you have finished pushing and patting down the pebbles or other aggregate you are adding, the slab should look like this—no trace of pebbles to be seen. Now wait for the surface of the concrete to begin to harden before hosing.

5. As soon as the concrete shows sign of initial set, test by hosing lightly and scrubbing with a broom. If this does not overexpose or dislodge the pebbles, you will know it is time to proceed with this treatment over the entire slab.

6. Closeup shows how a typical pebble concrete surface should look during hosing and scrubbing. Your aim during this process should be to bring all pebbles or other aggregate as fully into view as possible while keeping the surface as smooth as you can.

CHAPTER TEN

HOW TO COLOR CONCRETE

You can produce a colored concrete floor—or steps, sidewalk, or driveway—in any of three ways:

You can color the concrete when you mix it or order it precolored from a transit-mix supplier.

You can trowel in color during finishing.

Or you can color the surface of the concrete later, with special paint, dye, or stain.

One drawback to the use of integral color is that it is hard to predict the shade you'll get. The appearance of the concrete while wet is no clue. I have seen concrete that was poured bright green from which all trace of pigment had disappeared within a week.

The nearest thing to a rule is that full-strength pigment will usually produce a strong color when you use 7 pounds of color to each bag of cement. Never use more than 9 pounds. This assumes you are using pure mineral oxide, prepared especially for use in concrete. For pastel colors, use about 1½ pounds to the bag of cement.

White portland cement will give you cleaner, brighter colors. Use this rather than the normal gray cement except when making dark-gray, brown, or black concrete.

For uniformity you should mix the coloring material with the cement first, and mixing time for the concrete should be longer than usual for thorough blending. Measure all ingredients carefully—again for uniformity of appearance.

If you are using transit-mixed concrete, you are limited to the colors your supplier offers. But the job is greatly simplified. You simply place and finish the concrete just as if it were the plain gray stuff.

Incidentally, you can combine colored concrete with pebble topping for unusual and attractive effects. How to apply and expose the pebbles is described in the preceding chapter.

Because coloring is pretty expensive stuff you may prefer to use a two-course method, even though it involves a little more work. In this system you place a base course of ordinary concrete in the usual manner, roughly leveling it to about ½ to 1 inch below the top of the form. Allow this to stiffen

To give a concrete floor a general cleaning before painting or dyeing or staining, scrub it with one pound of trisodium phosphate in a gallon of warm water. To remove spots of oil or grease or paint, dissolve a pound of lye carefully in a gallon of water. Pour this on shavings or sawdust and let stand for several hours or overnight.

After soaking in lye, scrub stained spot with putty knife or old chisel. Repeat if necessary with stubborn stains. Lye solution can also be used to remove old paint, if you have the patience to do it. Or you can use a floor sander with open-coat silicon carbide paper, No. 3½-20 or 4-16.

After cleaning the floor, acid-etch it to remove slick surface and open the pores to allow stain or dye to penetrate. Mix 1 gallon of muriatic acid with 2 to 3 gallons of water, pouring the acid into the water. Use an enamelware crock or a wooden bucket. Mop or brush it on the floor and let it stand till the bubbling stops.

Work the acid into the floor, persisting till there are no slick spots. Then flush the area with plenty of clean water and let it dry thoroughly. You may have to wait a couple of weeks with a basement slab unless you provide special ventilation with an electric fan or otherwise.

until the surface water has disappeared. Then complete the pour with colored concrete made without coarse aggregate.

The process of striking off, floating, and troweling is the same with colored concrete as with any other. A light brushing after troweling is generally a good idea for concrete outdoors. For an indoor floor you may prefer to trowel a smooth finish. Cure carefully to avoid staining.

Usually the simplest and most reliable method, as well as the most economical, for making a colored slab is by troweling in pigment during finishing. This is sometimes called the dry-shake system.

With this method you mix and pour ordinary concrete and strike it off in the usual fashion. Then, again as usual, you level it further with a darby or a bull or hand float. A magnesium or aluminum float is best for this, used after any free water has evaporated from the surface.

Concrete dye should be poured into a flat-bottomed container and applied with a stiff brush. If some spots stay glossy or lighter colored than the rest of the area, scrub additional dye into them vigorously. Some dyes are one-coat processes. Others call for two coats, the second serving as a dressing.

Stain may be applied with a brush, spray, roller, or flat applicator. Smooth surfaces will require one coat, porous ones two with a 24-hour wait between coats. Widely available cement floor stain shown here is made in a variety of shades of gray, green, red, and brown. Follow it with application of paste wax.

Then it is time to shake the dry color material as evenly as you can, by hand, over the surface. The dry-shake material commonly consists of mineral oxide color blended with white portland cement and specially graded sand. This is offered ready to use by a number of manufacturers. It is not something you should try to compound yourself unless you are more interested in experimenting than in assured results.

By the way, either colored or plain gray concrete can be given a glittering surface by use of silicon-carbide grains. Spread and trowel these into the surface. These grains are highly reflective and glitter in sunlight or artificial light. You might like to use them on a driveway or sidewalk or footpath that is hard to illuminate adequately at night.

Give the dry-shake pigment a few minutes to take up some moisture from the concrete, and then float it thoroughly into the surface. Repeat, using about half as much dry-shake, floating in the same way. Then trowel at once.

When the surface has set up sufficiently, give it another steel-troweling. You can follow this with a third troweling if you want a very smooth and dense surface—as you might for a floor inside a house. For an exterior slab, which could be dangerous under damp conditions if too slick, a light brushing with a fine push broom is a better idea.

A colored slab needs the same curing as any other. After it has cured and the surface has dried you can enhance the appearance by giving it two or more coats of concrete floor wax containing the same pigment used in the dry-shake.

The many products available for adding color to an existing slab include dyes, stains, and paints. If a slab has been painted before, about the only thing you can do is paint it again. Otherwise, dyes and stains, which penetrate the surface and require less upkeep, are usually a better bet.

The pictures show how to dye or stain a slab, including the preliminary treatment necessary for a good job.

CHAPTER ELEVEN

SHORTCUTS AND BACKSAVERS

LET'S FACE it, the easy way to do concrete work hasn't been invented. But experience *has* produced a number of techniques to lighten the load.

Gathered in this chapter is a collection of tested techniques to ease the strain of home-style concrete jobs. Some will save you time. Some will save you money. And nearly all of them will pay off where it counts most—right in the middle of your otherwise-aching back.

1. Break up a big job into little ones. Concreting a whole width of porch or terrace at once can be a backbreaker even for two people. A series of 4 foot squares is a cinch even for one.

Using methods pictured and described in detail in the next chapter, build a grid of rotproof redwood 2-by-3's, then fill the squares with concrete. At a session you can do as little as a single square or as much as half the squares, checkerboard pattern.

If you want to keep the wood free from stains, varnish it before you pour the concrete.

2. Use timbers for forms when your concrete work is the first step in putting up a building or a porch. When you form with lumber you can reuse it later and have no waste at all. And heavy timbers will stay in place with minimum staking.

Hose and broom off cement from the timbers as soon as you pull the forms. Then they'll clean up easily for staining or painting.

3. Score lumber or plywood on the back when you need a sharply curved form. A series of saw kerfs about an inch apart and going two-thirds of the way through ⅜-inch plywood, for example, will make it limber enough to bend easily. Such cuts can be produced rapidly with a power saw, measuring by eye being accurate enough for placement of cuts.

Such cuts make the material flexible enough to take proper shape easily. It must then be braced by stakes driven into the ground to prevent shifting under the weight of wet concrete.

4. Use water pipe as screed down the center of a wide slab. It's tough for one man to strike off level a slab wider than 5 feet (for two men, 10 feet). A screed down the middle lets you pour and level off half the job at a time. After that you pull the screed and fill the hollow with concrete.

Pipe is straight, easily set, and leaves a smaller hollow than a 2-by-4 does when pulled out. It may be reused indefinitely in this way before being given its final job as part of a water line.

5. Cast stepping stones instead of a continuous sidewalk. It's far easier— and in most situations more attractive.

For one thing, you'll have only half as much concrete to handle. Small stepping stones of concrete can be much thinner than a big sidewalk without much chance of cracking.

The form work is easier too. And if you want irregular shapes you often can cast your stepping stones right in place. Just dig a series of 2-inch-deep, sharp-edged depressions in the ground and fill them with concrete.

6. Use steel reinforcing for added strength. Use of steel rods and welded wire mesh will permit you to reduce the thickness of your slab or wall by one-third to one-half. On even a moderate-sized job this can mean a reduction of thousands of pounds in weight you must handle.

If you must do a concrete job in a place that is difficult to reach with truck or even wheelbarrow, give even greater attention to designing for optimum use of reinforcing. It is far easier to carry in a few rods or a roll of wire than dozens of extra bucketfuls of heavy concrete.

7. Keep a watchful eye on the quality of mix as you do a concrete job. If you use a suitably rich mix with sand and aggregate proportioned as prescribed in the chapter on mixing, you'll be sure of strong concrete. If it's strong it won't have to be excessively thick at the expense of your back.

Remember that an overly wet mix is weak and that it is just as great a danger with transit-mixed concrete as with the kind you mix yourself. Transit-mix drivers are inclined to add too much water if you permit them to. They know this makes for easier pouring. But they will defer to you—it's your concrete, after all—if you insist on keeping the mix stiff enough for quality concrete.

8. Lift reinforcing into place as the cement is poured. Doing it even a few minutes late may be backbreaking—or impossible. And reinforcing that hugs the ground is wasted.

For normal slab and sidewalk conditions, pull the reinforcing up so that it is approximately centered between top and bottom of slab.

9. Use the dry-pack method when building a pool. This eliminates the big job of building forms and also lets you pour the whole pool in a single operation.

The dry-pack method calls for digging a rounded hole with sloping sides and then packing very stiff concrete against its sides and bottom. Although it is called dry-pack, the concrete is not actually dry—just sufficiently stiffer than usual so that it will stay in place against a slope.

For a big pool the packing is done with a shovel. For a small wading pool or garden pool you would place the concrete partly by dumping and partly with a shovel and then use a trowel to shape it.

The pool chapters tell in more detail how to use this time- and effort-saving method.

Share a load and save your back. This canvas sling with pipe handles spreads the weight of 2-foot-square concrete stepping stones, made with concrete left over from pouring a driveway. It is especially handy when heavy objects must be moved up steps or across a yard where a wheelbarrow won't go.

10. Use big tools for big jobs. When you try to do all the leveling of a slab after the striking-off process with a wood float—as backyard concreters are likely to do—you waste a lot of time and effort.

As you can see in the pictures of tools in an earlier chapter, both bull float and darby are simple devices that you can easily make from a couple of pieces of pine. Time spent making them will be saved in one use.

A darby is especially handy for a sidewalk or narrow driveway, all of which can be reached from outside the forms. It is also less awkward than a bull float when you are doing a slab surrounded by walls.

Use a bull float for quick leveling of a wide slab outdoors, such as a patio or a slab for a new building.

11. Hold up final troweling until all water has disappeared from the surface of the concrete.

Premature troweling is the biggest and most frequent of beginners' errors. It doubles the work, leads to wrinkled and mottled surfaces likely to scratch, gather dust, and develop hairline cracks.

For smooth results and minimum effort, hold up final smoothing with the steel trowel until using it briskly produces a ringing sound.

12. Broom-finish outdoor jobs instead of steel-troweling them unless you really need very smooth concrete.

A broomed surface is usually more attractive, always less slippery. It cuts out the most backbreaking part of the task by eliminating the final troweling.

Just wood-float the surface and then steel-trowel to reasonable smoothness. Depending on the texture you want, stroke with a kitchen broom of straw or plastic or with a hair-bearing push broom. With grids, alternate the direction of brooming for a checkerboard pattern.

And don't forget that an exposed-aggregate surface or a pebbled one is even more attractive and—since it eliminates fine troweling—less work. The textured and more colorful surface is better suited than ordinary concrete to most yards and gardens.

13. Add color to your concrete as you finish it. A paint, stain, or dye job later is a far bigger enterprise and may well have to be repeated again and again over the years.

The dust-on method described in the chapter on coloring adds little to the work, sometimes actually reduces it by easing the troweling.

14. Cure your concrete the easy way—but cure it thoroughly. A well-cured slab is a stronger one, so it can be thinner. This means economy in both materials and effort.

If you've taken up sod, you can form it into a dam to keep the new concrete flooded for the first week of its life. (Some soils will produce a slight staining.)

The other traditional methods are laborious—covering with earth, sand, straw or rags, or constant sprinkling. Simple alternative: spread low-cost polyethylene sheeting over the concrete within a few hours after finishing and let the slab provide its own moisture.

CHAPTER TWELVE

HOW TO REPAIR CONCRETE

THE MAN in the checked shirt is doing the two kinds of repair jobs you are most likely to encounter: patching stucco or filling a crack in a step, sidewalk, or slab.

The standard mix for repair jobs is one part cement, two parts sand, two parts pea gravel. For narrow cracks where pea gravel won't fit, use mortar consisting of one part portland cement to three parts sand. Or use dry-mix concrete or mortar—the kind that comes in sacks or small packages. In any case, make your mix stiff. The photographs and their captions tell how to do a typical repair job.

For difficult patching, especially on vertical surfaces, you can be much surer of permanent results if you use a concrete adhesive in addition. Some types can be used both on the surface to be patched and in the patching mix, as an additive.

One adhesive that has solved some sticky problems for me is Bondwell Acrylic Resin, a milky fluid made by the Franklin Glue Company. For a reliable patch, I first brush this onto the old concrete and wait for it to become tacky. Then I make the repair with mortar or stucco in which I have used some Bondwell in place of part of the water. This way I have a very plastic cohesive patching mix and also a sticky surface for it to adhere to.

Simplest material for repairing plaster or stucco is the dry-mix product sold in small sacks. Just mix it with water and trowel it on, following directions on the package.

To repair small cracks in concrete, use dry-mix mortar. For larger ones, you can use a concrete mix, following the steps shown in the next four photographs.

Use a chisel to roughen the area to be patched. Knock away any crumbling material that might eventually work loose. A wire brush, if you have one, is helpful in doing this. Use blown air or a water rinse to get rid of dust. The patch will hold better if you undercut the crack or edges of the hole slightly.

Dampen the concrete or stucco where the patch is to go. But don't be so liberal with water that there is any standing in the crack. If you're using a cement adhesive, substitute it in place of the water in this step.

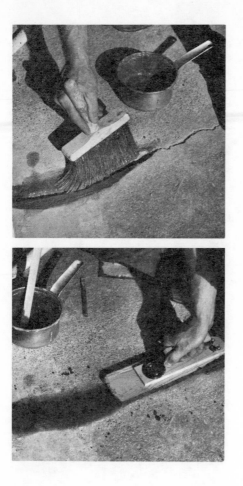

Prime the surface with a thin, creamy mixture of portland cement and water. If you have used a cement adhesive, you can omit this step.

Trowel or float the patching mix into the crack or hole. Press hard to force it into place, tamping if necessary. Be sure to fill the aperture at least level. Overfill slightly if possible to allow for slight shrinkage in drying. Cover the patch with vaportight material, such as a bit of polyethylene sheeting for at least five days so it won't dry too fast.

PART II

CONCRETE TECHNIQUES

AND PROJECTS

CHAPTER THIRTEEN

POUR A GRID PATIO

REDWOOD STRIPS or grids with pebble concrete between them make an almost ideal floor for an outdoor living area such as a patio or porch.

This combination is rapidly replacing the traditional approach of paving the whole place with a single dull expanse of troweled concrete. In the town where I live, a gasoline station using this treatment for its entire service area has been picking up architectural prizes.

When you divide an area with redwood strips and then texture the concrete you pour between them, you gain five ways:

1. You avoid the backbreaking and palm-blistering job of troweling the surface.

2. You divide your patio into smaller areas that can be thinner—hence cheaper and easier to pour—without danger of cracking.

3. You break up the job into small parts, so you need do only as much in a day as you feel like. What used to require a crew can be a one-man job.

4. A textured surface is safely nonslip, and relatively free from glare even in full sunshine.

5. And with the infinite variety of surface textures and grid patterns possible, you add decorative quality to the utility of a paved dining or lounging patio.

Early users of this method confined themselves to regular grids of redwood 2-by-4's, placed to form 4-foot squares. This arrangement is still good, especially for patios with curved outlines, but less formal patterns are often more attractive.

For the entrance patio of the home my family and I have built for ourselves in Carmel, California, I wanted a pattern that was both abstract and nicely balanced. So I went to the master of geometrical arrangements, the Dutch painter Piet Mondrian, by running through a book of reproductions of his work—something to be found in any fair-sized public library and in inexpensive paperback form in book and art stores.

I chose Mondrian's 1921 "Composition in Red, Yellow, and Blue." The reproduction in my paperback edition was about 3 by 5 inches and my patio space is 15 by 20 feet, making the scaling simple. It isn't necessary to be particularly precise about this anyway.

Any grade of redwood can be used for patio grids as long as it is one of

For a handsome porch like this, use rough 2-by-3 redwood strips and reinforce the concrete so it will support the cantilever.

Redwood 2-by-4s in a varied geometric pattern separate area of pebble concrete hosed lightly to expose some aggregate.

Concrete cast in strips is far easier to handle than a single large area. White cement makes interesting contrast with dark pebbles.

Large round pebbles thickly cast and fully exposed in a base of white concrete produce this effect. Grid is of redwood 2-by-4s.

This terrace is worth noting for unusual use of concrete block in wall, as well as grid pattern of 2-by-4 redwood pairs.

Not all grids need be wood. Here foot-wide sections of white pebble concrete divide large areas of textured gray.

Concrete surface of author's redwood-grid patio combines exposed aggregate with a relatively light seeding of pebbles.

the all-heart kinds. The sapwood parts of redwood may eventually decay, just as most other wood species would do. With both minimum price and maximum durability in mind, I used foundation-grade for my 2-by-4 grids. The grade called "construction heart" is also excellent.

If you have a couple of friendly volunteers on hand, you can pour the patio all at once, using transit-mixed concrete. A solitary worker can spread the job by pouring a few rectangles at a time, in a sort of checkerboard pattern for accessibility.

Air-entrained concrete is recommended if you live in a cold climate. It has a chemical in it that produces billions of microbubbles that allow for expansion and reduce the likelihood of cracking during a freeze. This kind of concrete is easier to work, too. For air-entrainment when mixing your own concrete, use cement labeled Type 1A.

If you are using transit-mixed concrete, ask the supplier for 5 to 7 percent air-entrainment. It is also well to ask for concrete that contains at least six sacks of cement in each cubic yard, for easy working and extra strength. If the question of consistency comes up, say "4" slump. This will give you a good, workable mix. Stiffer mixes are hard to finish by hand methods, and soupier ones don't give the strength and durability you want.

As soon as you have struck off the concrete level with the grids, strew flat pebbles on the surface and work them in with the strike-off board. If there are places where the concrete sets up too fast for this, lay a board over some of the pebbles and beat them down in. Details on doing this and on hosing off the concrete to expose the pebbles and some of the aggregate are covered in the chapter on pebble concrete.

After leveling subgrade, assemble and nail redwood grid, blocking up where necessary. Give the whole patio a slight slope for drainage. Rocks and chunks of old concrete can be tossed in, especially at low spots, to get rid of them and to save concrete.

If you wish to protect redwood color from graying effect of cement, coat top surfaces with clear plastic or varnish—and be gentle when striking off concrete. In most yards and gardens, however, the soft gray to which redwood weathers is most attractive.

Fill the grid pattern one rectangle at a time, and immediately strike off level. Then distribute pebbles at once and work them into the surface before the concrete becomes too stiff. If delayed you can still pound them in—but that's the hard way.

Except for curing, that finishes the job. As with any concrete, this kind should be kept damp for a few days. Simplest method, especially in a hot climate, is to lay polyethylene sheeting over the surface to keep the moisture in.

For the grid pattern of your own patio, you can choose your own Mondrian or adapt one of the styles shown in the photographs.

Because it is both handsome and rotproof, California redwood is the logical material for grids. Often a good alternative to the usual 2-by-4 redwood is rough 1-by-3 redwood—used for all grids or for a part of them. Since the rough stuff is a full inch in thickness, it is almost two-thirds as thick as surfaced 2 inches; and 1-by-3 will cost you only a little more than one-third as much as surfaced 2-by-4. The thinner material is especially suitable for a small porch or patio or for a pattern needing many divider strips.

Given stable soil or a good base layer, 3 inches is thick enough for patio paving when the rectangles are fairly small. This is another economy made possible by the grid plan.

Using a few pebbles—preferably quite large flat ones—will give you quite a different effect from using many or using small ones. You can also get one kind of texture by hosing soon so that much aggregate is exposed and another by waiting so long that the concrete between pebbles remains nearly smooth.

If you find you've waited too long to bring out the hose, increase the pressure by nozzle adjustment. If necessary, scrub away some of the surface cement with a stiff broom as you spray.

But precision in timing and expertness in technique really aren't terribly important. The wonderful thing about building a patio this way is that even if the result isn't just what you had in mind, it'll be handsome all the same.

Properly shod—boots in cold weather, nothing in warm—you should test the surface with a hose as soon as it appears partially set. If too much cement washes away, wait awhile. If pebbles or aggregate are difficult to expose, scrub with a broom.

CHAPTER FOURTEEN

BUILD A DRY-PACK SWIMMING POOL

OF THE many ways to construct a swimming pool, the one most dependable in nonprofessional hands is the dry-pack method. This involves digging a spoon-shaped depression in the ground and then lining it with reinforced concrete. There are no forms to build and tear down with all the waste of lumber involved, and there is no business of pouring a floor on one occasion and walls on another with the accompanying problem of avoiding a leaky crack where the two pours join.

The pictures show a pool that novelist and longtime do-it-yourselfer Paul Corey and I built by the dry-pack method. Not counting the excavating, we and our families and friends put 118 man-woman-and-child-hours into it over a period of nine days.

It's a pretty spacious pool, 53 feet long and 23 feet wide at its widest, and holds 20,000 gallons as nearly as we can estimate. Since it is not intended for diving, its greatest depth of just under 6 feet is adequate.

Cost of materials for a pool like this is about $400, hardly one-tenth of what a pool of such size ordinarily costs when contractor-built. Our figure is for just the pool, of course; it does not include fencing, surrounding patio work, or a filter system. Most of the $400 goes for 1,000 square feet of welded wire mesh, 1,200 feet of ⅜-inch reinforcing rod, and concrete. We used 18 yards of sand and gravel and 84 sacks of cement.

If your pool location, unlike ours, is easily accessible for a transit-mix truck, you can save a good deal of labor and probably a few dollars in cost. On a job this size, discuss your requirements carefully with an expert at the transit-mix plant before ordering. You'll want a comparatively rich mix with at least six sacks of cement to the cubic yard. And you'll want it to arrive stiff enough to pack on the sloping walls of your excavation without any tendency to slump.

If you mix your own you'll quickly learn how dry to make it. Although the general formula for dry-pack concrete calls for 1 part cement to 2½ of sand and 3½ of gravel, you may find a slightly richer mix (1 to 2 to 3) is more easily worked. Mix each batch just a little longer than you might normally do.

For a job this size you really should round up enough friends to make a

The dry-pack pool, made of $400 worth of concrete and reinforcing steel, is 53 feet long, big enough for quite a crowd.

four-man crew, or even a five-man so that one can be resting. Figure on two to mix and dump into the wheelbarrow, one to wheel, and one to spread.

With transit-mixed concrete, the two on the mixer are eliminated. But a four-man crew is still preferable and will get the job done twice as fast as when having to mix as well as place. With four men and two wheelbarrows you can discharge the truck fast enough to avoid excessive stand-by time, which is expensive.

Roll each wheelbarrow load to the edge and dump it over, to be spread with shovel and rake. It is important to watch the preplaced steel reinforcing during this spreading, making sure that all of it is pulled to a position near the center of the concrete. Rake and smooth the concrete to a thickness of about 6 inches at the bottom of the pool, 4 inches on the sides. Use shovels to pat it into a compact and reasonably smooth mass.

After the dry-pack concrete has been placed and taken its initial set, keep it wet for at least three days for strong curing. You shouldn't do this by filling it with water because the concrete isn't strong enough yet to carry the load. Frequent sprinkling is one way, but polyethylene film to hold in the existing moisture is easier and better.

75

For appearance you'll probably want to put some kind of coping around the edge of the pool. Flagstones set in a bed of mortar are good.

If you'd like a gleaming white pool, a trick taught me by a professional pool builder will be useful. It's a white coat applied with sacking, and it gives a slightly rough surface that is safer than a slick one with a pool that has sloping sides. You can put it on after a few days.

Mix 1 part of white cement with 2 parts white sand and enough water to make a heavy paste. For a pool the size of ours you'll need about two sacks of cement and four of sand. Apply this white coat by dumping a shovelful of the pasty mixture onto the dampened surface of the concrete and spread it around with a pad of burlap. Rub it in vigorously to fill all hollows and rough spots. Replace the burlap as it wears out. Go back over the job after a little while the rub off the rough spots.

Pick-and-shovel is the slow way to excavate for a pool—but it's cheap, and the only way in locations inaccessible to power equipment.

Carpenter's level on a camera tripod can be sighted along to act as a crude transit. Use it in making pool rim level.

Line the excavation with reinforcing steel. This pool required 1,000 square feet of welded wire mesh and 1,200 feet of ⅜-inch rod.

Help keep reinforcing in position with twists of baling wire. Rocks hold steel up in places where it might be trampled.

Entire pool should slope to a drain. Unless you have a generous supply of water and other use for it, plan on a filter system.

Dry-pack pour begins with dumping of rich, stiff, well-mixed concrete over the pool rim, to let it spill down the sides.

After concrete has been dumped over the side of the pool excavation, it must be raked to proper thickness.

Short length of reinforcing steel can be bent to form a handy tool for pulling reinforcing to position at center of concrete.

Pack the concrete by whacking it with the back of a shovel. The more you whack, the better the finish.

Set coping in mortar all around the pool rim. Scraping with shovel knocks off spilled mortar and any rough spots in concrete.

White coat smooths and whitens rough surface left by dry-pack job done with shovel. Surface must be damp for good adhesion.

Excessive roughness in white coat can be removed the next day by light rubbing with emery cloth or piece of old sander belt.

White-coated pool looks like this as initial filling with water begins. Gap in coping is overflow so water can carry off leaves.

Poolside walkway must slope away slightly to carry off rain water. Concrete poured between coping and bank needs no forms.

Safety fence to keep wandering children out is a necessity.

CHAPTER FIFTEEN

POUR A WADING POOL

THIS CHAPTER is a sort of postscript to the preceding ones. Its pictures show how to adapt the techniques for building a swimming pool to the much simpler project of casting a smaller one for kids to wade in.

Actually, when the dry-pack method is used, such a pool can be big enough to interest adults too, in hot weather, and still be only a one-weekend project.

When you consider that a pool like this will last almost forever, it must be classed as one of the biggest bargains that $25 has ever bought.

The photographs show the major steps in building a wading pool. There is additional information on this and other small pools in the next chapter.

Partial coping of stones at far end of pool is an optional extra. It increases pool capacity, adds decorative touch, and partially conceals water supply. This is an ordinary hose bibb overhanging pool, which is also used for garden hose.

Just as with big pool, job begins with free-form, spoon-shaped excavation. Some kind of waterpipe drain should be provided.

Stiff concrete—see preceding chapter—is dumped along pool sides and shovel-packed against them. Use of reinforcing mesh is desirable but, with a small pool that is poured 4 to 6 inches thick, is not essential if well tamped earth base is provided.

Although pool may be used as soon as cured, later painting will make it more attractive. As described in the chapter on coloring concrete, this process begins with a scrubbing and an etching with acid to provide a good surface for the paint.

Swimming-pool paint should be used. Painter's assistant works ahead of him with vacuum cleaner to keep surface clean.

CHAPTER SIXTEEN

GARDEN POOLS

A GARDEN pool will charm the eye and do much to make outdoor living areas seem cooler on hot days. Its basic ingredients are a hole in the ground and a little concrete.

A reflecting pool for purely decorative purposes need hold only a few inches of water. One for fish and plants should be from a foot to 2 feet deep, but it can be rough and rustic. A wading pool should be from 1 to 1½ feet deep with smooth sides to protect small feet and make it easy to clean.

To lay out straight-sided shapes, drive stakes at the corners and run string between them. You can mark out a right-angle corner accurately by using that carpenter's friend, the 3:4:5 ratio. That is, if you measure 3 feet from the corner one way and 4 feet the other and the distance between these two points is 5 feet, you have a precise right angle.

To mark out a circular pool, drive a stake at the proposed center, loop string over a nail in the top of the stake, and swing the string like a compass, driving stakes as you go. For a free form, just toss down a length of garden hose or a rope and move it about until it suits your fancy.

You can set the whole pool below ground level, or build it like a well housing, mostly above ground. If the pool is to have vertical sides, begin by excavating to approximate size. Where soil drains poorly or winters are cold, dig 6 inches deeper than pool depth and fill with 4 to 6 inches of gravel or broken rock, tamped hard.

A fish pond should rarely need refilling, but a wading pool may have to be cleaned out every day during the hot season. A water-supply line saves the trouble of handling a hose. But be sure that its opening is well above the highest level at which water can stand in the pool. Having the opening below the surface might permit contaminated water to be drawn back into the line.

Draining is simple when the surrounding ground is lower than the floor of the pool. You need only install a drain and enough pipe to lead waste water wherever you wish. Or you can omit the drain and use a siphon or jet pump to lift the water over the edge of the pool. But if the pool is in a low spot, you must either pump it out or put in a drain running to a dry well. This may be made as in the drawing.

The simplest kind of drain is a galvanized-pipe elbow set, open end upward, flush with what will be the surface of the pool floor, and at its lowest point. A standard bathtub stopper can be used to plug a 1¼-inch elbow. In a fish or garden pond, which will rarely be drained, you can close the elbow with a pipe plug, or provide an overflow by screwing in a suitable length of pipe. To drain the pool, all you will have to do is unscrew the pipe.

Pour and spread concrete evenly to half the floor thickness desired. Then place ⅜-inch reinforcing bars or heavy mesh on it and pour the other half, to a total of 4-inch thickness in warm climates, 6 inches where freezing occurs. For a sloping pool, that's all you do.

Vertical-wall pools are a bit harder. One way to make the walls is to lay up stone, brick, or blocks with mortar, as for a building. Coating them inside with a rich cement mixture will make them hold water. Brick can also be laid dry, without mortar between. Lay one or two courses at a time, a few inches inside the excavation walls. Pour concrete between the brick and earth walls every other course or so. With smooth brick, this method can be used even for wading pools.

BUILDING AN INSIDE FORM

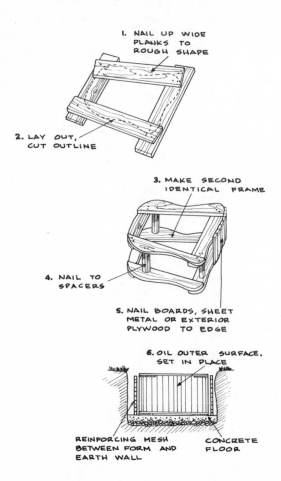

1. NAIL UP WIDE PLANKS TO ROUGH SHAPE

2. LAY OUT, CUT OUTLINE

3. MAKE SECOND IDENTICAL FRAME

4. NAIL TO SPACERS

5. NAIL BOARDS, SHEET METAL OR EXTERIOR PLYWOOD TO EDGE

6. OIL OUTER SURFACE. SET IN PLACE

REINFORCING MESH BETWEEN FORM AND EARTH WALL

CONCRETE FLOOR

Follow these steps in casting a curved or free-form pool with vertical sides. Earth sides of the excavation make the outside form, and concrete is poured between earth and inside form.

Plumbing for your garden or wading pool can be quite simple. Drain the pool by gravity to lower ground if possible; otherwise build a dry well as shown in the middle sketch. An overflow pipe screwed in as in the bottom sketch is an excellent device. It controls the water level at all times and also allows you to drain the pool at intervals without groping around for a plug somewhere underwater.

GRAVITY DRAIN
TO LOWER GROUND

BARREL   LID

DRY WELL   STONES

OPEN BOTTOM

OVERFLOW PIPE

UNSCREW TO
DRAIN POOL

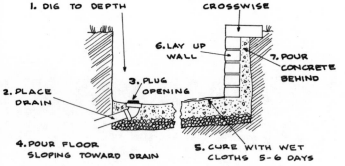

1. DIG TO DEPTH

8. LAY COPING CROSSWISE

6. LAY UP WALL

7. POUR CONCRETE BEHIND

3. PLUG OPENING

2. PLACE DRAIN

4. POUR FLOOR SLOPING TOWARD DRAIN

5. CURE WITH WET CLOTHS 5-6 DAYS

To build a sunken pool lined with brick or stone or block, follow these steps. The wall can be laid up with mortar or it can be stacked dry, with concrete then poured between it and the sides of the excavation.

You can easily build an above-ground pool by laying up brick or stone walls on a concrete slab, using mortar. In cold climates, make the excavation deep enough to take a 6-inch layer of gravel in addition to the 4 to 6 inches of slab.

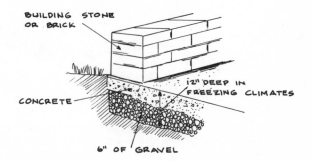

BUILDING STONE OR BRICK

12" DEEP IN FREEZING CLIMATES

CONCRETE

6" OF GRAVEL

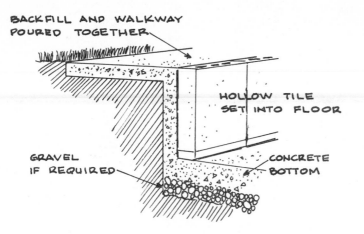

BACKFILL AND WALKWAY
POURED TOGETHER

HOLLOW TILE
SET INTO FLOOR

GRAVEL
IF REQUIRED

CONCRETE
BOTTOM

A big garden pool or wading pool for children is very quickly made by using hollow tile. Dig a straight-sided excavation 2 inches deeper than the length of the tile units and 4 inches bigger all around than the pool is to be. Mark out the walkway and dig it 4 inches deep. Pour the pool floor of concrete and stick in the tile units at once. Trowel the floor smooth, let it cure for several days, then backfill with concrete between tile and excavation. At the same time, pour the concrete for the walk.

To add movement to your garden pool, build in a fountain or waterfall, using a submersible pump. Textured wall of this pool was made by carefully removing forms within a few hours after the concrete was poured and then strongly spraying the surface with a garden hose.

Surrounding brick for this pool matches patio surface of common brick laid in sand. The pool itself is simply a shallow pan of poured concrete.

Poured concrete pool takes its character from sharply squared-off coping and unusual staggered planter, both made of concrete block.

Shallow pool of poured concrete is brought to life by use of a tiny immersible electric pump. Low free-form wall of brick in mortar separates pool from patio and sandpile.

Like the wading pool in the preceding chapter, this garden pool was made by packing comparatively dry concrete against the bottom and sides of a spoon-shaped excavation.

CHAPTER SEVENTEEN

WALLS AND STEPS

SOLID CONCRETE garden and retaining walls are not difficult to pour, but they do require forms that are strong and fairly elaborate. To build the forms, follow the tips and the drawing in Chapter 2.

The trouble with an ordinary concrete wall is that, like a troweled slab, it is a little too industrial looking to harmonize with garden surroundings. There are two simple ways to improve such a wall: insert rocks during the pour, and expose the aggregate afterwards. Either of these tricks will help a lot. Used in combination, they produce a handsome and rugged wall that looks as if it has been there always—and will last just about that long.

The techniques required are covered in detail and shown in photographs in the chapter on how to pour a fireplace. Briefly, you should acquire a collection of rocks, each having one comparatively flat face. As you pour the concrete, insert rocks into the forms so that the flat face of the rock is against the wood of the form. To give more texture to the wall, pull the forms as soon as it is safe to do so—usually a matter of two to four hours. Then hose down the surface vigorously with plenty of water, just as when making an exposed-aggregate slab. This is described in Chapter 8.

You can avoid all the form work involved in poured-concrete walls by using concrete block. For your garden wall of block, begin by casting a generous footing. Usually you can pour your concrete right into the excavation without building wooden forms. The bottom of this footing should be on firm soil at least 18 inches below ground level. Since it should extend below frost line, you will have to go even deeper in colder areas. A general rule is that the footing should be twice as wide as the thickness of the block wall and half as deep as it is wide. So for a wall of 8 inch block, your footing should be 16 inches wide and 8 inches deep.

The simplest way to start the block part of the wall is by placing the first course of block right in the top of the poured footing while it is still wet. Level the block right away, of course. You need place only this first course of block on the day of the footing pour.

If you don't find it possible to work fast enough to place the block right in the concrete, just level the top of the footing. Then, when you are ready to build the wall, place the first course in a full bed of mortar spread on top of the footing. The pictures in Chapter 20 show how to lay block walls.

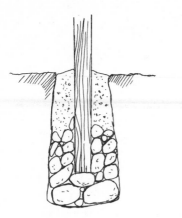

Even where your plans call for a wooden fence rather than a masonry wall, concrete can add solidity and permanence to the job. Dig a hole for each post about 5 inches deeper than you plan to set the post, making it about 4 inches wider than the post at the top and 8 inches wider than the post at the bottom. Set with rocks and concrete as shown. Using a level is the easiest way to make sure the post is plumb, but if you don't have one at hand there's another method. Just sight the post against a couple of house corners, if at all possible; they're pretty sure to be properly vertical.

For a substantial garden wall, begin by casting a concrete footing on firm soil at least 18 inches below ground level—below frost line if that's deeper. Make the footing twice the wall width and half as deep as it is wide. Then build the wall, placing the first course of block in a full mortar bed on the footing. If you are able to place that first course while the footing is still soft, you can save time and mortar by just wiggling each block right into the concrete.

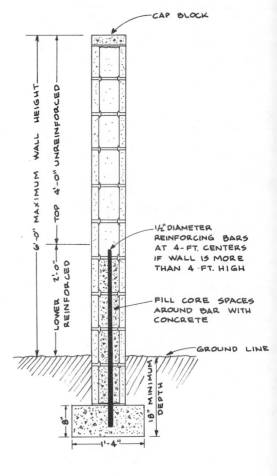

CAP BLOCK

6'-0" MAXIMUM WALL HEIGHT

TOP 4'-0" UNREINFORCED

LOWER 2'-0" REINFORCED

1/2" DIAMETER REINFORCING BARS AT 4-FT. CENTERS IF WALL IS MORE THAN 4-FT. HIGH

FILL CORE SPACES AROUND BAR WITH CONCRETE

GROUND LINE

18" MINIMUM DEPTH

8"

1'-4"

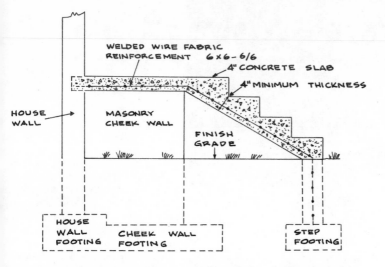

WELDED WIRE FABRIC
REINFORCEMENT 6 x 6 - 6/6

4" CONCRETE SLAB

4" MINIMUM THICKNESS

HOUSE WALL

MASONRY CHEEK WALL

FINISH GRADE

HOUSE WALL FOOTING

CHEEK WALL FOOTING

STEP FOOTING

This section drawing shows how a typical porch floor and steps may be cast in a single concrete pour, using continuous mesh reinforcing.

Here is one simple way to prepare the forms for pouring concrete steps and porch. Since a comfortable height for a step is 7½ inches, you can build these forms with a minimum of carpentry by using dressed 8-inch lumber, which will be close to the desired width. Unless steps are quite short, it is best to use 2-by-8 rather than 1-by-8 lumber, to avoid bulging when the concrete is poured. With a 7½-inch riser, a good tread width for steps is 10 inches or a little more.

This section drawing shows how to build exceptionally attractive steps on sloping ground that is a little too steep for a ramp. The redwood forms become a permanent part of the steps, and a few nails are left sticking out to key the lumber to the concrete. The plan at left shows how such steps as these can be given an especially pleasant and informal look by staggering them slightly.

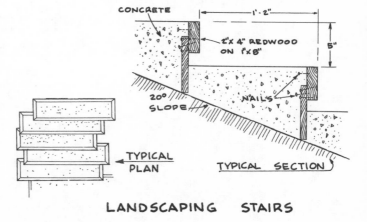

CONCRETE

1'·2"

2"x 4" REDWOOD ON 1"x 8"

5"

20° SLOPE

NAILS

TYPICAL PLAN

TYPICAL SECTION

LANDSCAPING STAIRS

There are two ways to build a garden wall or planter like this one. You can use concrete block laid in what is called stacked bond, with joints not staggered. Or you can pour concrete into forms to which narrow sticks have been tacked to produce a pattern of lines. Redwood seat topping the planter wall is a nice added touch, useful and inviting.

Curved forms being comparatively difficult to build, a wall like this is best built of concrete block. An assortment of block sizes, all of the type called slump block or slump stone, was used here. Wall is topped with a thin layer of mortar to provide a smoothly finished surface.

Continue your block wall to the top and then cap it with cap block or brick or flagstones to close off the hollows. With a low wall, it is often simplest just to pour the block wall full of concrete.

If you need unusual strength or if your wall is to be more than 4 feet high, reinforce at least the bottom 2 feet of it with steel and concrete. To do this, set ½-inch reinforcing bars well down into the footing at intervals of 4 feet. Be sure they're placed so the block will fit over them. Usually if you place the first bar properly the 4-foot interval will take care of this.

Then after you have laid the first 2 feet of your wall, fill the spaces around the bars with concrete. This will give you a sort of pilaster of reinforced concrete every 4 feet. Even for a 6 foot-high garden wall, the bars and concrete usually need extend up only 2 feet.

BUILDING STEPS. The number-one rule for building steps is: don't make them too steep. A standard figure is treads 10 inches wide with 7½-inch risers. For short flights and changes of level in a yard or garden, you'll want to make the steps broader and give them somewhat less rise as well.

In conventional step-building, footings for the side walls are placed on firm soil below frost line. Side walls are tied in with the foundation walls with anchor bolts or tie rods. Crushed stone or gravel is used to make a fill before the concrete is placed; this fill should be well compacted by tamping. The concrete is then placed in two layers—a base course and a wearing surface. This permits making both the treads and risers smooth.

1. These three photographs show the procedure for building standard steps. Note that crude though these forms appear they are actually straight, accurate, and thoroughly braced. Fill forms like these with a stiff-mix base course and tamp it, allowing 1 inch for wearing surface.

2. Place the fine concrete for the wearing surface within 45 minutes after the base course is tamped. Starting at the top, remove the riser forms one at a time as soon as the wearing surface has been placed and finished. Immediately plaster riser face, using the wearing-surface concrete mix.

3. Unless troweling is required to match an existing finish, avoid using a steel trowel on step treads. A wood float, as seen here, should be used to produce a gritty, safe, non-skid surface. Use an outside edger to round the corner of each tread and an inside edger, if you have one, at the bottom of each riser.

Textured concrete combines with use of rocks to produce an informal effect suited to a garden setting.

Rough 4-by-6 redwood is riser and part of tread for each of these steps, and acts as form as well. Nails driven at varying angles will key with concrete when it hardens, keeping wood and concrete permanently together. Leading edges of 4-by-6s have been rounded slightly. Concrete work consists merely of shoveling in the mud—a special mix made with rounded rocks and tinted a soft brown—leveling it, then hosing it off after an hour or so to expose the aggregate. With less concrete required and no forms to pull off later, this system makes it easier to produce unusually stylish steps than it is to pour a flight of ordinary all-concrete ones by traditional methods.

Concrete for the base course can be composed of 1 part portland cement to 2¼ parts sand and 3 parts gravel or crushed stone. It should be comparatively dry and stiff. For the wearing surface, use 1 part cement to 1½ or 2 parts each of sand and gravel. Don't use any aggregate of more than ⅜ inch for the wearing surface. Place and finish the concrete as shown in the photographs.

By introducing some modifications into this basic procedure you can create steps that are far better looking than standard ones and more suited to home, as opposed to industrial, settings. Such steps are easier to build too. Use the same techniques suggested for walls earlier in this chapter. Both exposed-aggregate and pebble-concrete finishes are fine for steps, and safer than a slick finish besides.

Still another excellent trick—and my favorite—is an adaptation of the procedure used in casting a grid patio. In this one you use heavy redwood for the crosspieces of your forms and leave this redwood permanently in place. Such steps are highly decorative, especially when you then hose off the concrete to expose the aggregate or pebbles pressed into the surface. One of the photographs shows an arrangement for doing steps this way.

CHAPTER EIGHTEEN

DRIVEWAYS, WALKS, AND STEPPING STONES

SINCE A sidewalk or a driveway is just a special kind of slab, you can build an excellent one by following the principles already covered.

Minimum width for a driveway should be 8 feet, and more width is a great convenience. For a double garage, width at the doorway will be 16 feet, and if the driveway is not too long this width should be maintained all the way. A car can then be left in the driveway without blocking the other half of the garage.

A conservative recommendation for thickness of an ordinary driveway is 6 inches. In practice, however, many are only about 4 inches thick—the usual thickness for sidewalks—and serve nicely if reinforced with wire mesh. Contraction joints at 10-foot intervals (and down the center of a wide driveway to a double garage) are a wise precaution.

If you prefer to build a wheel-track driveway of two strips, make them about 24 inches wide, with a 3-foot space between them.

The standard driveway or sidewalk finish is either rough-floated or lightly troweled and then broomed for texture. Earlier chapters tell how to produce these finishes.

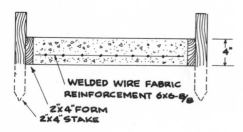

This section of a concrete sidewalk shows how simply one may be poured by using forms made of standard 2-by-4 lumber—for both sides and stakes.

WELDED WIRE FABRIC REINFORCEMENT 6×6-6/6

2"×6" FORM

2"× 4" STAKE

Same kind of formwork as used for a sidewalk makes a driveway when the scale is enlarged, as shown here. Where dip in center is wanted for drainage, it is made by using a strikeoff board that has been sawed to this profile.

A form arrangement like this is useful for casting thin paving blocks and stepping stones or thick blocks to be laid up as steps. Blocks for steps will usually be at least 6-inches thick and so heavy you should plan to move them as short a distance as possible. Stepping stones can be 2 to 3 inches thick. Surfaced 1-by-6 or 1-by-3 lumber makes suitable forms for blocks and stepping stones. Wiping forms with crankcase oil will make stripping and reuse easier.

This much pebble-concrete driveway is a lot of work. But so is any normally finished, and charmless, driveway of this size. The chapter on pebble concrete tells how to give any slab this look.

A large expanse of concrete is actually easier to produce when done in two stages with contrasting finishes. Colored concrete is interestingly combined here with the pebble variety.

Contrasting areas of rough-floated and exposed-aggregate concrete are separated by a grid of 2-by-4 redwood. The zigzag of the grid effectively echoes the shape of the fence.

When an expanse of concrete is wide, the grid divider need not be wood strips. Here it is white concrete strewn with large, flat pebbles. Main areas have a coarsely broomed finish.

To harmonize concrete with lawn and garden, use divider strips of grass. Curved forms, later replaced by soil, are made by laminating three or four thicknesses of 1-by-4 lumber.

You can easily build a distinctive sidewalk like this one with precast concrete blocks, purchased or homemade. Place blocks to a stretched string, press pebbles into mortar joints.

Making contraction joints is the only technique you may need in sidewalk and driveway work that is not usual with other slabs. Use a straightedged board to guide the tool.

For a driveway or sidewalk that is more than utilitarian, introduce texture or pattern or both. The photographs show many ways in which this has been done by use of grids or exposed-aggregate or pebble concrete—or by combinations of these. The most important ingredient of a modern driveway or sidewalk is, it turns out, not the concrete but a lively imagination.

A final tip: not all walkways need to be paths of uniform width. Often a series of stepping stones made of cast concrete will serve the foot as well and do a good deal more for the eye.

One of the two basic methods of making a stepping-stone path begins with the casting of a collection of concrete rectangles, preferably of varying size. Making these is a good way to salvage excess concrete when doing other work with transit mix. The rectangles are then moved into place and arranged in a pleasing and convenient pattern.

Even simpler are poured-in-place stepping stones. To make a path of this kind, carve out 3-inch-deep depressions in the ground and fill them with concrete.

For either kind of stepping stone, the easiest finish is exposed aggregate. If bare feet are to use the path, however, be sure the aggregate is of the round kind and that it is only lightly exposed to avoid excessive roughness. Much easier to walk on barefoot is pebble concrete, preferably the kind made by adding only a comparatively few pebbles and allowing the concrete surface to become quite stiff before hosing it off just enough to expose the pebbles.

103

CHAPTER NINETEEN

PAVE YOUR YARD THE EASY WAY

BY PAVING a section of your yard you can multiply its usefulness and cut maintenance way down. Children will be grateful, and so will you.

There is a hard way to pave a yard, and there is an easy way. By choosing the latter you can give your yard the comfort and convenience and good looks of garden paths and terraces for outdoor living. And you can do it without backbreaking effort.

The approach that lets you do this is not the standard procedure for concrete work, not by a long shot. Compared with the usual methods it will cut your time and effort to less than half. And it will save you more than fifty cents out of each dollar.

It is a method I finally came to after using up a good many yards of fill and sand and gravel, a good many sacks of cement, and a considerable total of sweat doing paving the traditional way.

Mind you, I'm not quarreling with the standard specifications for concrete work. Paving must often be done to specifications drawn in advance and figured to meet all known weather conditions. A driveway that may occasionally carry several tons of truck must be thick and well based.

But there's no need to apply these standards and requirements to a garden path to walk on or a paved terrace where you plan to put nothing heavier than a chair or table.

Sure, there are other problems. Tree roots can crack concrete. Frost in the ground can heave and break it. The standard answer to this—and it doesn't always work, either—is a lot of fill plus thick concrete. But let's take a look at the four elements that make standard concrete work costly and laborious. And let's see how this system gets around all of them.

These four things are: leveling the ground; putting in fill; building forms heavy enough to be level and rigid; mixing and pouring—or buying and placing—enough concrete to make a slab 4 inches thick.

Suppose now that, instead of going against nature and fighting her with heavy layers of concrete, you go along with her.

To begin with, *don't* level the ground. Let paths and terraces follow the natural contours, doing just enough shovel work to smooth out bumps. Human

Paving a chunk of yard creates a play area and improves circulation—in this case among sandbox at left, wading pool at right, lawn in foreground, and house beyond.

If yard has a natural slope, concrete can follow it—for easy form building and better drainage. Forms and grid strips are ½-by-2-inch redwood. Planks at right are for wheelbarrow.

bulldozing is frequently half the job, so eliminate all that. A terrace with a fair amount of pitch to it is not much good for putting a billiard table on, but it will do nicely for sitting. And rain will run off it in a hurry.

Don't put in heavy forms, rigidly staked. Use curves most of the time instead of straight lines. A curved member is likely to be many times stronger than a straight one, besides being more pleasing to the eye, so your forms can be ultralight and still need only occasional staking down. This means pennies instead of dollars spent on form lumber.

Fairly stiff mix of rich concrete is best for relatively thin concrete layer like this. Where forms cannot be used, as at poolside here, some leveling has to be done by eye.

Pleasant texture is produced by dragging a household broom lightly across the concrete. This gives a low-glare, nonskid surface that is easy to clean and not too rough for bare feet.

Here is an optional extra at practically no added cost: Paint on a chess or checker board or provision for any other sidewalk games favored by family members.

Don't use any fill, unless your soil drains badly or is otherwise notably unreliable, because fill is expensive and adds a lot of work. How do you get away with this? Well, that's a big part of the secret of low-effort backyard paving.

Instead of making continuous walks and big slabs, convert your whole job into a series of simple stepping stones.

This, as you can see by looking at the photographs, does not change its appearance or usefulness at all. But it does change the way you do the work, making it much easier and cheaper and faster.

Break up all areas of concrete into many smaller ones by using divider strips of wood to create a series of rectangles. For sidewalk these can average 2 by 3 feet, for the terrace 3 feet or so square.

The result: if heaving occurs from roots or freezing, each rectangle of concrete will move independently. If it is lifted by natural forces, it can settle back later (or be put back, if necessary) without a crack forming. You have made concrete into a flexible material—by going along with nature instead of fighting her.

The wood strips also act as expansion strips. All the form lumber can be rough 1-by-2's, easy and cheap to buy. But if fairly sharp curves are wanted, the lumber must be thinner. For the walk-and-terrace job you see in the pictures, I used half-inch lumber of the cheapest grade offered and was able to get quite strong curves. I used redwood since it will last a long time without rotting out.

And finally (to get that fourth point covered)—*don't* pour a slab 4 inches thick. You can get away with making it only half that thick because your job consists of such small units.

107

So there is the final economy: only one-half or so of the usual amount of cement and gravel to buy, only half the mixing and carrying and spreading to do.

After your first section of concrete has had a day or two to start curing, you can rip off the outside forms and reuse the lumber for the next section. And because the job is being done in a series of rectangles, it is easy to stop at any time without leaving an ugly joint. Several short sessions in the cool of early morning or late afternoon are much easier on the back, and possibly the heart, than an all-day-long splurge of mixing and pouring and finishing. And when only half the squares in a terrace are done in a day, it is easy to reach each of them for finishing without stepping on fresh concrete.

Even in the final step you can avoid almost entirely one of the bugaboos of amateur cement workers: troweling to a fine finish. This happens to be one of those jobs that you cannot do well without a good deal of experience, and even then it's a backbuster. Just follow one of the procedures outlined in the chapters on textured surfaces and pebble concrete. For instance, a broomed surface, used on the yard-paving job in the photographs, is not only easier to produce than a slick finish but is more attractive to the eye and practically skidproof to boot.

PART III

WORKING WITH MASONRY

CHAPTER TWENTY

BLOCK IS WHAT YOU MAKE IT

THE ALMOST limitless variety of patterns, colors, and textures possible with modern concrete-block units make this suddenly one of the newest and most excitingly versatile of all building materials. You can use block most successfully for retaining walls, garden walls, house walls, fireplaces, barbecues, and—as shown in a later chapter—patio paving.

The handsome wall effects developed recently depend upon color and pattern. For the first of these you can choose blocks with color cast into them, or you can use paint—including new rubber-based masonry varieties. Pattern comes from choice of patterned or textured blocks or from the way you lay them—running bond, stacked, staggered or on end—or how you tool the joints.

Blocks made from ordinary aggregates are heavy, like standard concrete. For most home jobs, from housebuilding to patio walls, you'll prefer the lightweight variety. They are easier to lift and place, give better insulation and sound-deadening.

Most common size is 16 inches long, 8 inches high, 8 inches wide. (Blocks actually are ⅜ inch shorter and lower than these nominal dimensions, to allow for a mortar joint of that thickness.) Use these blocks in the usual way, staggering the joints in what is called a running bond, and you get the kind of wall so often seen in industrial construction. Such a wall is so ugly that, in my opinion, it should never be used for any purpose where it will be visible.

A better answer is to use blocks that are 4 inches high instead of 8 inches. Their bricklike shape, also familiar in adobe, is much more pleasing.

Some makers now offer a foot-square block. This is usually stacked with one block right over the one below, giving mortar joints that run from top to bottom of the wall.

Blocks may also be had in the size as well as the shape of common brick. Unlike the larger block, these usually are solid rather than hollow.

Thin, flat blocks are usually intended for paving jobs—to floor a patio or path. But there are irregular thin blocks for wall use too, where they give an effect rather like stone.

For extra texture you have a choice between split block and slump block. The split form has a rough face but in regular dimensions, usually long and thin. Slump block is a peculiar offering with a remarkably unblocklike appear-

Ingenuity creates a charming wall of ordinary block turned, in extraordinary fashion, on its side. Cutting block to make mortar-lined opening is pretty tedious, of course . . . but what a result! Unusual paving in foreground is redwood rounds bedded in sand, with concrete filling spaces between.

Here's a way to combine any degree you wish of privacy and ventilation, using only a single type of block and its matching half block. Simply place some units normally, others on their sides. Instead of using all 8-by-8-by-16-inch units you can, for economy, mix 4-inch-high blocks on their sides with 8-inch high and 4-inch-thick ones in normal position.

Even walls with such extreme textural qualities as this one can be built with standard block units—when mixed with imagination. Concrete bricks outline the pool below. The wall at left is made of split block laid in what is called "stacked" bond—giving continuous vertical joints.

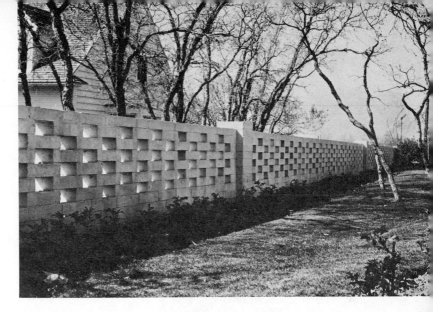

Spacing is all that is needed to produce openness and ventilation in a block fence. Top course in this one is the same kind of block as the rest, but mortar joints have been made level with block surface to give the effect of a continuous beam topping the open block wall.

Here is another example of how joint treatment can effect the whole character of a wall. Most mortar joints in this wall were raked very lightly to let them almost disappear, but every third vertical joint was raked heavily to produce a sharp shadow line. Note how block gives effect of adobe.

A good lesson in handling change of level in a long block fence is offered by this wall of split block. Note also that the horizontal has been stressed by tooling horizontal mortar joints while striking off the vertical ones flush with the face of the block to make them virtually vanish.

Use of split block makes this planter far more attractive than one built of ordinary smooth units. Best choices for any wall needing a rugged texture are either split block or slump block, seen in another photograph. Split block is especially good where painting is not contemplated.

Slump block are produced by using a special mix and removing forms early enough to allow some sagging of the concrete. They are good-looking painted or unpainted and are the nearest thing possible to the traditional adobe of the Southwest. Note how this wall extends from indoors to outdoors.

ance. It is block made from such a concrete mix that the units sag or slump when taken from the molds. Both height and texture vary enough to give an informal and rugged look to a wall. It is among the handsomest of blocks.

The most striking patterns of all are those you'll get if you use any of the new patterned blocks. Several varieties are shown in the photographs accompanying this chapter.

But plain blocks can produce almost equally interesting patterns. A simple change from a running bond to vertical stacking is enough to convert a commonplace wall into a distinctive one. Also possible, and highly effective, is a pattern made by laying two standard 8-by-16-inch blocks one above the other, then the next two on their ends. This gives the kind of checkerboard often seen in brick floors.

Scoring the blocks is another way to vary the pattern. Do this by running a masonry saw across an ordinary block to add a shadow line. On walls that are not to be painted, you can make this score line match the regular joints by filling it with mortar, then treating it just as you do the joints.

Combining two types of block in one wall creates a whole new set of possible patterns. You can alternate between an 8-inch-high block and a pair of 4-inchers, one above the other, in each course—or in alternate courses. Or you can simply lay a course of high blocks, then a course of half-height ones, and so on. It's easy to work out variations on this scheme to get many of the patterns called ashlar, popular in stone work.

Treatment of the mortar joint has an equally great effect on pattern. The normal thing is called a tooled joint. Obtain this by letting the mortar partially harden, then smooth and compress it into a half-round cove with a tool which may be a short length of steel rod or pipe bent into a sharp S-curve.

That's standard—but it isn't inevitable. The joint may simply be rubbed fairly smooth and made almost invisible. Use burlap to rub it, at the same time cleaning excess mortar off the face of the block.

As striking as any pattern, particularly on a wall that gets the sun, is the squeezed-joint. To get this, merely use plenty of soft mortar as you lay the block, let it squeeze out from the joint—and leave it alone.

Another treatment for a sharp shadow pattern is raking, in which the mortar is cleaned from the joint to a standard depth of half an inch or less. This throws the shape of the block into sharp relief.

A combination of joint treatments multiplies the possibilities. It also permits some interesting camouflages. Lay a wall in the standard way, then wipe the vertical joints flush and smooth while raking or tooling the horizontal ones. When such a wall is painted, the up-and-down joints disappear and the effect is surprisingly similar to wood siding or clapboard. It's a successful trick for making a house look longer than it is. And it will help a concrete-block garage, for instance, harmonize with a frame house.

Low-maintenance has always been a popular feature of block. You can reduce the upkeep even further by avoiding paint. Treat the wall instead with a silicone waterproofing liquid. The natural gray of ordinary block may be just what you want. If not, the new colored block, most often offered in soft greens and tans, is an answer.

Each of these remarkably various and handsome screen walls is built in precisely the same way. On a standard footing of poured concrete the decorative units are set in mortar in the stacked-bond pattern that produces continuous mortar joints vertically as well as horizontally.

For small color modifications on either natural or tinted block, you can add a little color to the silicone waterproofing, converting it into a waterproofing stain. For more vivid color, inside and out, use one of the paints made especially for use on masonry. They're now coming in a wide variety of brighter hues.

The old reliable for exterior use is portland-cement paint. It comes as a powder—basically cement plus pigment—to mix with water and apply to a wall that has first been dampened with a fog spray from a hose nozzle.

A variation of this, for texture rather than color, is a wash of cement and water. Use this on a wall made of blocks 4 inches high by 16 inches long and you'll have something that closely resembles adobe.

The old idea that concrete-block walls are damp and chilly has been pretty much knocked out by the use of lightweight blocks. When such a wall is insulated by pouring loose fill into the hollows, its heat loss is about 20 percent less than for standard wood-frame construction with plaster. To insulate the wall this way, just pour rock wool or vermiculite into the hollows directly from the bag it comes in. Insulating fill of this kind will also support the concrete of a bond beam—see next chapter—making metal lath unnecessary.

Block walls do an efficient job of cutting down noise. Because they are comparatively heavy they do better than most wall materials at reducing sound *transmission.* Most types of block walls 6 inches to 8 inches thick give a reduction factor of 40 decibels or more even when not plastered. This puts them into the "very good" range. Normal speech is not audible through them, and loud speech can be heard faintly but not understood.

Sound *absorbing* is often more important than transmission. This governs how much the noise within a room is bounced about. Where hard plaster or glass absorb only about 3 percent of this noise, masonry units soak up as much as 68 percent. Best are lightweight aggregate blocks with a coarse, open texture. Painting reduces the absorption somewhat, but thin, sprayed coats do so less than heavy ones.

CHAPTER TWENTY-ONE

HOW TO BUILD WITH BLOCK

FOR MANY walls, especially those using the modern decorative blocks shown in the preceding chapter, you'll need only a single kind of block unit. Other walls, such as those that are part of a house or fireplace, call for a few special-purpose units. These supplement the ordinary, or stretcher, blocks, the ones that make up the bulk of any wall.

Six or seven types of special-purpose block are in fairly common use.

Half blocks are just what they sound like—units one-half the length of full stretcher blocks. Whenever a wall laid in the common running bond comes to an end or an opening, one of these is called for.

Corner blocks have one end cast flat instead of having a half-hollow at each end. Blocks of some manufacturers, however, are all made with flat ends— and with these no special type is required for a corner.

Double-corner blocks are similar but have both ends flat. Use them for piers or columns that are only a single block long.

Bull-nose blocks are corner blocks with one corner rounded. You may want them for openings or corners even if you are using blocks of a pattern having flat ends in the normal units.

Jamb blocks are useful for doorways.

Casement-window blocks have slots down the ends to take steel window units.

Channel blocks are formed like troughs. Lay them end-to-end to form a continuous opening. Pour concrete into this channel, around reinforcing steel, and you have a bond beam.

If you possibly can, decide on the block size before you design your wall. It's a waste of time to make a fireplace some odd length, or to set a door in some arbitrary place, and then find that you have to chip or cut off blocks to fit.

LAYING A BLOCK WALL. A block wall should sit on a footing of poured concrete. For a fireplace, indoor or outdoor, this will ordinarily be a large concrete pad. For a regular wall, the footing will be poured concrete that goes below the frost line.

There's a simple formula for footing size. Make the footing as deep, and twice as wide, as the wall is thick. If you are using 8-inch-thick blocks, your footing should be 8 inches deep and 16 inches wide. Then the block wall will be set in 4 inches from the edge of the footing both front and back.

If you are putting a wall where there is a drainage problem, place drain tile along the outer edge. Slope it about 1 inch for each 20 feet, cover the tops of the joints with roofing felt, and backfill with 12 inches of gravel or crushed stone. An additional answer to this problem is given in the next chapter.

If the blocks you're going to lay have to be stored outdoors, keep them covered for protection from rain. Wet blocks will shrink on drying and give you weak—and possibly leaky—joints.

Make your mortar by mixing two shovelfuls of masonry cement (or one shovelful each of portland cement and hydrated lime) to four to six shovelfuls of mortar sand. Mix with just enough water to give a plastic mortar that clings nicely to trowel and block without being soft enough to squeeze down too much when you lay the block. You'll learn the right consistency very quickly as you work.

Start laying block by spreading a bed of mortar on top of the footing. As you put in each block after the first, butter one end of it with mortar, using a mason's pointed trowel. Or, as some do, butter the end of the preceding block instead. Or both. Whichever works out simplest for you will give good results.

Squeeze the block against the preceding one to give a ⅜-inch joint. Use your trowel to cut off any excess mortar that oozes out—unless you want to let it harden that way for the looks of it.

When laying any course after the first, spread mortar on top of the laid block in the area the new one will cover. Usual practice is to put this mortar only along the front and back edges.

Unless you're doing a job where irregularities are okay, check each block with a level as you go. Check to be sure it is level both ways and that you are getting a plumb wall. You should have a string stretched from one end of the wall to the other.

A wall of brick-shaped or slump blocks is often more interesting if it is imperfect. Inaccuracies frequently add to the charm of garden masonry. So don't use level and string too assiduously where perfection is not called for.

Mortar will remain usable for about two hours in hot weather, three in cool. Any not used in that time should be thrown away. Within that time you can retemper your mortar by working it over, adding a little water when needed.

As the mortar joints stiffen up, go back and wipe any spilled mortar from the face of the blocks with a gunny sack. Then tool the joints according to the pattern you desire.

TOPPING A WALL. The top course of an exposed wall of hollow-block should be made solid to keep out rain. You can buy solid blocks for this purpose. Or you can set strips of expanded metal lath in the last horizontal joint. With this to support it, use mortar or concrete to fill the voids in the blocks of the top course.

A good way to pull the wall tightly together is to make the top course a bond beam. Use this method on block fences and also for walls of buildings.

Use channel blocks with metal lath under them, and fill the trough they form with concrete. Adding two lengths of reinforcing steel will greatly increase the strength.

You can use this same method to construct a lintel over a window or doorway. Support it rigidly till the concrete sets. Or give it the permanent support of a pair of steel angles under it. Even easier—use precast lintels if available.

If your wall is to be topped with a wooden plate (normally two 2-by-6's, one on the other) to support rafters, you'll have to set anchor bolts. These should be ½-inch bolts 18 inches long, placed not more than 4 feet apart.

Where each bolt is to go, place metal lath 16-inches below the top of the wall and fill the hollow with concrete to hold the bolt. The bolt must protrude above the block wall by a little more than the thickness of the plate you are to use.

STRENGTHENING A WALL. There are several ways to make a block wall even stronger. You may need them when building a pier or pillar or perhaps a retaining wall. They're recommended in earthquake country too and where winds are violent.

One is to use extra-strength mortar. Make it by using only two or three shovelfuls of sand to one of portland cement, and omit lime or use only a quarter of a shovelful.

And you can use what are called full-bed joints. You get these by spreading mortar over the entire top of the block instead of just along front and back.

You can get even more strength by reinforcing the wall with steel and concrete. Do this horizontally by pouring a bond-beam midway up the wall and at the top. For vertical reinforcing, drop a length (two at the corners and ends) of reinforcing steel down a hollow every 4 feet and fill the hollow with concrete, keeping the steel in the center. This is most effective if the steel overlaps another piece of steel that was inserted into the footing when it was poured and allowed to stick up about 10 inches.

The photographs show wall-building procedure step by step, from laying out the block to determine position to capping the finished wall. By adapting the techniques shown you can build just about any kind of block wall you want—and do a professional job of it.

1. Before you place any mortar, string out the block on the footing to check the fit. A mark on the footing at each block joint will help later to keep the spacing correct.

2. Spread mortar for the first few blocks at one corner. Furrow it with a trowel to make sure there will be plenty of mortar along the bottom edges of the face shells of the block.

3. Place the corner block solidly in the mortar in correct position. Check it both lengthwise and crosswise with a level before going on to lay additional blocks.

4. Using a mason's trowel, butter the ends of the face shells of each block before laying. Buttering with mortar is easier if the block is held in a vertical position.

5. After laying three or four block, use a mason's level or a straight board to check alignment. Hold it against the outside edge of the block as well as along the top.

6. Make sure that block are plumb as well, tapping each block with trowel handle if necessary to change its position slightly. Lay all block with thicker end of face shell up.

7. For courses after the first, place mortar for horizontal joints along face shells of block already laid. Butter and place each new block just as you did for the first course.

8. Build the corners of the wall first, usually four or five courses higher than the center. Except for low, informal walls where precision is unimportant, keep checking alignment and plumb.

9. Measure height and length of wall occasionally to make sure that mortar joints are staying close to the standard ⅜''. A board held diagonally along block corners gives another check.

10. When filling in the wall between corners, string a line for each course at the height of the top of the block. Hold each block as shown and roll and shove it into position.

11. Make any adjustments immediately, while mortar is still plastic. Moving the block after the mortar has stiffened will break the mortar bond. This is an important rule.

12. Use your trowel to cut off excess mortar soon after laying each block. You can use this mortar again by tossing it back onto the board and working it into the fresh mortar.

13. When mortar has stiffened sufficiently, compact the joints with a tool like this or the S-shaped one seen in the next photograph. Tool horizontal joints, then vertical.

14. Tool shown previously is best for horizontal joints, while this one is for vertical—but one like this will serve well enough for both. Bend it from ⅝-inch bar or tubing.

15. After tooling tye joints, remove mortar burrs with a trowel or by rubbing with a burlap bag. Use the burlap also to remove spilled mortar that may be clinging to the wall.

16. A building wall, or a garden wall to be framed in wood, will need anchor bolts to hold a wood plate. Metal lath placed two courses down holds mortar around ½-by-18-inch bolts at 4-foot intervals.

17. Where bearing walls intersect, use metal tiebars like this, spaced not over 4 feet apart. Embed the ends in cores filled with mortar or concrete supported by metal lath.

18. You can tie in a nonbearing block wall like this one for a partition by using metal lath or ¼-inch-mesh galvanized hardware cloth. Use one strip in every second course.

19. Foundation walls need topping to distribute loads and bar termites. Other walls need top to keep out water. One way to cap a wall is by using solid-top block shown here.

20. You can cap a foundation or other wall without special block by using the method shown here. Cover the core spaces under the last course with metal lath strips in the mortar.

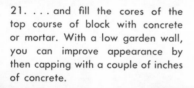

21. . . . and fill the cores of the top course of block with concrete or mortar. With a low garden wall, you can improve appearance by then capping with a couple of inches of concrete.

22. Or cap your wall with solid block units chosen and placed so joints break properly. With these, as with any block not having webs, butter the mortar onto the entire end of the block.

CHAPTER TWENTY-TWO

HOW TO WATERPROOF
A BLOCK WALL

BASEMENT WALLS and below-grade walls of cabins or other buildings should be given a coating on the outside to discourage water penetration.

The simplest method is to trowel on a ½-inch-thick coat of either portland-cement plaster or the mortar used in laying up block. This should be done on a wall that has been cleaned and dampened—but not soaked—with a water spray. The plaster coat should be kept moist for at least two days to make sure it is properly cured.

1. Clean and dampen wall and trowel on ½-inch coating of mortar, finishing it as smooth as possible. Cure by keeping moist for at least 48 hours. Extend this plaster coat 6 inches above grade.

2. For an even better job, make the coating only about ¼ inch thick. Then roughen it with a scratcher to insure a good bond. Keep it damp for at least 24 hours before applying a second coat.

3. Have surface damp, but not soaked, when applying second ¼ inch coat of mortar. Moist-cure this for at least 48 hours. You can follow this treatment with two grout coats if you need them.

4. Where wall meets footing, make a cove by sloping each mortar coat. This will help to prevent water from collecting around the juncture and eventually working its way through.

A similar, but better, procedure is to apply two coats, each about a quarter inch thick, scratching the surface between coats. The photographs show how this is done, and also how to form a water-shedding cove at the bottom of the block wall. This procedure, called parging, will provide sufficient protection for most basements. But if your soil is wet and poorly drained, you should add two more waterproofing coats. For each of these use a grout made of equal parts of portland cement and fine sand, mixed with water to the consistency of heavy cream. Dampen the surface each time and scrub the grout well in with an ordinary floor scrub brush. Cure each grout coat by keeping it damp for at least twenty-four hours. Instead of mixing your own grout, you can use cement-based paints sold for waterproofing.

Unless you are in a dry climate or have well-drained subsoil, you should also put in drain tile to carry off water. Place it around the outside of the footing and connect it to a suitable outlet. The joints between the tiles should be protected with bits of building felt and the tile should be covered with at least a foot of coarse gravel or crushed stone before backfilling is done.

CHAPTER TWENTY-THREE

SHORTCUT TO
FLAGSTONE PAVING

AMONG THE finest terraces and patios you'll ever see are flagstone ones. They are handsome, durable—and expensive. The cost of such terraces lies not so much in the materials used as in the amount of skilled time that goes into them. Conventional flagstone laying is costly to hire and slow and difficult to do.

And therein lies the great virtue of the dry method. It takes all the haste out of the job and cuts the work way down. Since it is so simple that anyone can do it without previous experience, it permits the luxury of a flagstone terrace or patio at small cost. You can use precisely the same method with clay or patio tile, sometimes called pavers, to produce a more formal terrace or porch floor.

Briefly, the shortcut method consists of covering a level area with sand, arranging the flags on it, sweeping over them a mixture of sand and cement, then sprinkling with a hose. It's as simple as that.

Any pleasant spot that can be made fairly level will do. It may be a place near your house, for dining or just sitting. It may be an area shaded by trees or wind-protected by the house, or one around a barbecue or picnic table. The same method can be used, of course, to build a walk or garden path.

After making the area approximately flat, cover it with sand. Use whatever kind of sand is cheapest and most readily available; ordinary coarse sand sold for making concrete works nicely. The sand should be at least 2 inches deep so that it will be easy to push the flags down into it.

For any climate where severe freezing may be expected, it is best to shape the earth so that water will drain off rather than stand and freeze. A bed of cinders several inches deep should go down before the sand, or the sand can be made 4 to 6 inches deep.

Rake the sand out smooth or pull a long board over it. Don't make it dead level; give it an inch or two of pitch for drainage. The terrace shown in the photographs was simply leveled by eye as the sand was raked out. More precise level can be obtained by staking a pair of 2-by-4's, one on each side of the terrace. The sand can then be brought to a level about 2 inches lower than the 2-by-4's. A board placed across them can also be used to maintain the level when placing the flagstones.

Flagstones picked up at no cost in an abandoned quarry and fitted without cutting produced this informal country terrace. For a more formal porch or terrace, use larger, smoother flags.

Any kind of flagstones will do. One beauty of this dry method is that it permits use of the less perfect—and hence less expensive—stones that are often available. Even large flat rocks can be used.

Fit the flags together to cover the sand as completely as possible. But maintain a gap of at least half an inch where the stones come closest to touching each other. Wiggle and work them down into the sand so that the flat tops are as level as possible. Push sand under any high points so that the stones are firmly supported.

Now wet down the flags and the sand between them. Do this far enough ahead so that the surfaces by the stones will have time to dry before you place the cement.

Mix portland cement with sand. For a fairly rough surface use coarse concrete sand, mixing one shovelful of cement to each three of sand. For smoother results use a fine sharp sand such as is sold for plaster work. Mix it one part cement to two parts sand.

If you'd like to have white or a pastel color between the flags, use white sand and white cement. These are widely available since they are used in making mortar for laying glass block. The mix for the terrace in the pictures consisted of two bags of concrete color. This produces a very attractive pale green.

Mix the ingredients dry. Use no water. Mixing can be done with hoe or shovel in a washtub or wheelbarrow. Or more easily in a concrete mixer.

Dump the thoroughly mixed sand and cement onto the flagstones. Sweep

133

this dry mortar across with a push broom or ordinary broom until all the gaps between the flags are filled level. Then brush the flags reasonably clean, using a broom or brush. A shop bench brush is handy for this but anything that will sweep will do.

The final step is the easiest of all. It consists of wetting the area with a fine spray from a garden hose. Do this until the mortar mixture has taken up all the water it can, but stop before much water stands on the surface. After perhaps fifteen minutes do this again. Repeat once or twice more so that the mortar becomes wet all the way down.

Then just leave it alone. No troweling or other hand work is needed. Come-back the next day and you'll find yourself with a fine flagstone terrace, made at a fraction of the usual labor and cost. For strongest mortar, keep the area damp for several days by covering with plastic film or by sprinkling frequently.

1. Begin by leveling the ground, but leave some pitch for drainage. In this well-delineated space between vacation house and pool curb, no forms were needed for leveling.

2. Rake the area smooth and then soak and tamp any loose fill to give a firm base. Dump and spread 2'' or more of sand to form the bed into which you will set flagstones.

3. Even such crude stones as these, fitted loosely without cutting, make a quick and handsome paving job when the dry-mix method is used. Coarse concrete sand forms the bed.

4. Dump the sand-cement dry mixture onto the flagstones and spread it with a broom. Continue sweeping the mixture across the flags until all the gaps are filled level.

5. Use a soft broom or a large paintbrush to clean the dry mortar mix off the flags. Then wet down the area with a fine spray from a garden hose. Repeat several times.

CHAPTER TWENTY-FOUR

LAY A SLATE HEARTH OR FLOOR

ONE THING that makes slate unbeatable as a do-it-yourself material is that it offers you a chance to save nearly 90 percent of the usual installed cost.

For floors and hearths, indoors and out, natural slate makes a dramatic, luxurious floor. It lasts forever, needs no maintenance except mopping, and stands up to anything from kids' toys to fire and flood. It should—at $4 or more a square foot, which is the going price in my area when laid by hired talent. But you can bring it off at about fifty cents a foot when you lay it yourself. And that's the average cost of an ordinary, nondurable floor.

Slate is a first choice for an impervious entryway, a hearth in front of a regular fireplace or under a prefab one, a floor for a porch or a room. I've used it for all those things in my own home, including floors for dining room and kitchen. It makes a good surface for a window-seat, too, where plants can safely be placed without danger of the moisture from watering harming impervious slate.

The slate you will find for sale at your patio- or building-supply dealer will probably come, as ours on the West Coast does, from Vermont. Of the two color types, gray-green and blue-gray, I prefer the greenish. Besides seeming softer and friendlier, it definitely shows dust and footprints less.

The usual way to lay slate is in a bed of cement mortar about ¾-inch thick. Since that's a good way and also cheap, I would recommend it for any large area you are going to cover. Over concrete it is the best way even for small jobs—unless thickness is important.

To the surprise of people who have done a lot of slate work, however, I have proved that slate also goes down nicely in mastic. This is an especially good way to do the work over a wood or plywood floor or where it is important not to add a lot of thickness. This is necessary if you are putting an entryway or hearth over an existing floor. Principal drawback to mastic is that it takes quite a bit of it to produce a level job when the slate is irregular in thickness. This adds enough to the cost to rule it out for big jobs.

Having made your basic decisions—where, how big, color, method—you are ready to buy your slate. For a small job you'll save time in the long run by choosing the pieces individually for easy fitting. Begin with straight-edged pieces for the edges of the area. Choose the rest for approximate fit.

Slate is easily extended from indoors to out, since it works so well in both places. Here it is used to pave both the entryway and front porch of a home built by the author in Northern California.

If you are using the mastic method, select slate for thickness as well. Choose pieces of average thickness—about ¼ inch—and avoid the occasional ones that taper noticeably.

Although many wallboard mastics will do, the safest choice is ceramic-tile cement, at $4 to $7 a gallon. A gallon covers about 50 square feet; much less if slate is irregular.

On a big job, such as a porch, you will find that you can save if you order your slate by the ton. This runs, in my area, $180 compared with ten cents a pound for smaller amounts. A rather generous estimate is 5 pounds for each square foot to be covered.

Surprisingly, the biggest part of slate installation is selecting the pieces and arranging them on the floor. The idea is to get fairly uniform mortar joints with a minimum of cutting.

Don't strive for too much regularity. Shapes that vary and joints that range from ¼ inch to as much as 1 inch give a slate floor the informality that is a great part of its appeal.

To mark the slate for cutting, slip one piece under the other and scribe with any pointed tool. Cutting is often done with a cut-off wheel mounted in a circular saw, but I prefer a slate nibbler. You may be able to rent one. They cost around $25.

When you've arranged and cut the pieces for about 40 square feet, mark around the slate with a heavy pencil or crayon. Then move the slate out of your way.

SLATE IN MORTAR. To lay slate in mortar over a concrete slab, mix an ordinary rich mortar of two or three shovelfuls of sand to one of portland cement. For a small job the convenience of packaged dry-mix mortar is worth the extra

Surprisingly neat fit of these pieces is easily obtained by use of a slate nibbler. Note how attractively this dining-room slate contrasts with the carpeting of the living room.

cost. Mix and chop with a hoe, using enough water to produce a smoothly plastic mortar.

With the surface of the concrete dampened, toss a trowelful of mortar where the first piece of slate is to go. Spread it to a thickness of about ¾ inch.

Plop your hunk of slate down onto the bed of mortar. Check it two ways with a level, pushing down as necessary. Make adjustments in level by placing a block of wood on the slate and swatting the block with a hammer. Or tap the slate with a rubber mallet or the rubber grip on the handle of an all-steel hammer. Proceed in the same way with the other pieces of slate, leveling each one to those around it.

When you've completed the area, level off the mortar between the slate pieces. A putty knife is good. Give the mortar an hour or two to stiffen and then smooth it off more accurately. Clean up spilled mortar as well as you can without disturbing the joints, but save final clean-up until several hours later. It will be much easier to do then, but it will still be pretty tedious. Find a helper, preferably female and skilled at cleaning floors, if you can. Any mess you miss can be removed next day with a steel-wool kitchen pad.

If you pour a concrete slab that you plan later to top with slate, just strike it off to an approximate level and leave it rough. The mortar will key better that way. If you are slating an existing slab that has been troweled smooth, you should begin by roughening it with a hammer and chisel.

Most of the above procedure for slate work with mortar applies equally to laying slate in adhesive. With the slate cut, position marked, and stacked aside, spread mastic on the clean, dry concrete or wooden floor or subfloor. If any slate is thinner than the rest, build up a thicker layer of mastic for it.

It's best to hold the mastic coat back from the edges of each piece of slate. This will allow space for the mortar to key better when you grout.

For the small amount required for grouting, it's simplest to buy dry-mix mortar. Stir it with water to the consistency of very heavy cream. Dampen the joint spaces with a sponge and pour in the mortar. Poke with a putty knife to make sure the mortar fills under the edges of the slate. Then proceed with the smoothing and clean-up as outlined above.

SLATE DRESSING. The floor will be usable by the next day, but the slate will have a somewhat whitish look. Mopping with plenty of clean water will help this. To make the floor look even better, mop it, let it dry, and then apply a slate dressing.

You can buy this in several types. I used a heavy, or varnish, variety on my first hearth. Until it wore off over the months, this gave altogether too much shine for my taste. So now I stick to the penetrating type that you can't even see; it just gives the slate a clean and pleasant glow. The kind I use is called Hornlux (A. C. Horn Companies, Atlanta, Chicago, etc.) and is sold nationally for use on terrazo. You just brush or wipe it on generously, let it soak in for fifteen minutes, then wipe off as much as you can. Other than mopping when dirty, this treatment occasionally is all the care a slate floor needs. That's just one of the reasons I think you'll like yours.

Playroom slate entry in same house as other two pictures was laid in mastic to keep it close to height of existing cork floor. Mortar was used to lay in the other rooms.

1. Begin by laying out the slate, choosing pieces for the best fit possible before cutting. Use any sharp tool to mark for cutting, allowing for mortar joints of at least ¼ inch.

2. Trim slate to marks. This slate nibbler is a hinged cutter in slotted steel. It lets you make curves and inside-corner cuts as easily as you can make the straight ones.

3. You'll need to round and polish only the edges of slate that will be exposed where a hearth or section of floor ends. Do it by hand with emery cloth or with a power sander.

4. With the location of each piece marked and slate put aside, spread mastic on the clean, dry concrete or wood floor. Or trowel mortar onto rough, damp concrete to make bed ¾-inch thick.

5. When using mastic to lay slate, you must add mortar between slate pieces to fill gaps. When you lay slate in a bed of mortar, more than enough will usually work up around the slate to fill joints.

6. Use a putty knife, pointing trowel, or similar tool to cut off and smooth the mortar to the level of the slate. Smooth again after an hour or two, and clean away any excess mortar.

CHAPTER TWENTY-FIVE

PATIO, TERRACE OR WALK LAID IN SAND

HERE IS the quickest way to build a masonry patio, terrace, sidewalk, or garden path or to put a hard-surface floor under a clothesline or in a play or service area. Just place precast masonry units, such as brick or concrete block—either purchased or homemade—in a bed of sand.

After marking out the boundaries of your project with stakes and string, excavate to give sufficient depth for 2 inches of sand plus the thickness of the brick or block.

Place your trim, the edging that will hold everything in place. This might be lengths of redwood 1-by-4 or 2-by-4, either rough or surfaced. Or you can use masonry units, possibly the same kind as for the paving, setting them on edge so that they key into the ground well. Wood edging should be held by stakes driven into the ground every 4 feet or so. Arrange this edging to provide a slope of at least 1 inch in 16 feet for drainage.

Spread sand over the area and, with the edging as a guide, use a notched board to level it to such height that your paving blocks will come to the top of the edging.

Start at a corner and lay brick or block to the pattern you have chosen. Use a trowel to wedge block into position. If a paving unit is too high or too low, just pull it out and take away a little sand or sprinkle more.

When you have laid all your block, sprinkle loose sand over the surface and sweep it back and forth until all the cracks are filled. Clean off the surplus with a hose.

Quickest and simplest patio you can build is one like this. Square concrete blocks 2 inches thick are set in a bed of sand. You could save by casting your own blocks, add interesting texture by giving them a pebbled or exposed-aggregate finish.

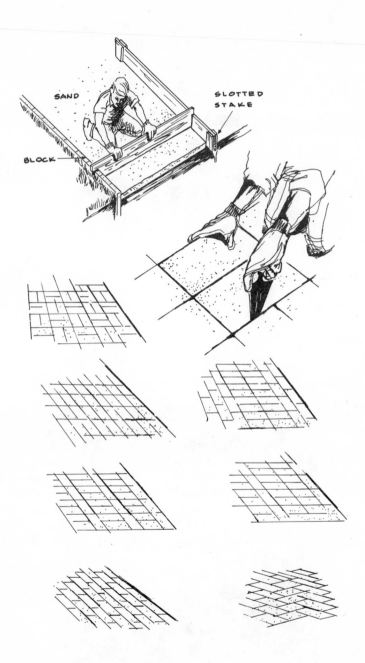

Seven possible patterns for a brick patio or walk and the two principal steps in creating one are sketched here. After laying out the boundaries and smoothing the soil, add sand to a depth of about 2 inches. Slotted stake and notched board shown are helpful in bringing the sand to an accurate level at the right depth for the brick or block you are using. A trowel is helpful in laying the block when it is used as sketched here.

This concrete-block-in-sand terrace gets a nice accent from use of a few blocks of slightly contrasting shade. Similar blocks set in mortar form the trim where a planter was called for. This two-and-two pattern is one of the easiest to lay.

Pattern is easily added to a block-in-sand terrace by using divider strips of wood. Or, as seen here, by getting the same effect even more easily from a change of color in the blocks employed. Concrete masonry units now come in many shades.

Terrace units, especially large ones like these, can be surrounded by something other than sand. Easiest way to produce this handsome effect is by casting the paving blocks in place, removing forms, then using joint spaces for planting.

Redwood grid must be built accurately to size if it is to be filled with common brick, as here, or concrete-masonry paving units. Just lay the bricks in a sand bed of proper height, pushing them tightly together, then sweep sand into cracks.

CHAPTER TWENTY-SIX

HOW TO FASTEN THINGS TO MASONRY

MASONRY WALLS and slab floors are tough to fasten things to. But there are standard answers to the problem that will save you accidents and headaches. Backaches too.

A fellow I know took along a rather unusual present when he went to a housewarming in a masonry home recently. What he gift-wrapped and fetched along was a pound of assorted concrete nails. These things are so hard and tough that they can be driven through structural steel or into concrete or blocks. His forty-five cent present, he was told later, was the best thing anybody brought.

These fat nails, ranging in length from ½ inch to 4 inches, are one of the devices that make it possible and easy to fasten even very heavy things to any kind of masonry and be sure they will stay fastened.

Around any house or garden where there is stone or block or concrete you'll have need of one or more of these systems frequently. And you'll need them even more if you do any remodeling or improving in the basement of almost any house. All of them have their uses on floors and walls alike.

Each method has different advantages.

Concrete nails have the merit of requiring no special tools, just the nails and a hammer, preferably a heavy one. You can drive them into hard substances, including even first-grade concrete that has hardened over the years, pretty much as you would hammer ordinary nails into pine.

You can use them to advantage around concrete construction. They will fasten corner bead, metal lath, stucco netting. They will hold furring strips to masonry walls that you may want to panel or nail dry-wall materials to.

Hard nails also offer the easiest way of putting down a floor plate if, for instance, you want to partition off part of your basement for a workshop or darkroom. Simply nail a length of 2-by-4 to the concrete floor.

Spring-wound drills offer a method of fastening to concrete more easily than in the past. These are carbide-tipped drills for use in an ordinary electric drill. Like other carbide-tipped drills for power use, they are much faster than hand work with star drills. An advantage of this type is that the spring winding pulls the powdered concrete out of the hole as you go. This means easier and faster drilling with no stopping to blow yourself an eyeful of dust.

146

Once the hole is drilled, fastening is accomplished by use of lead or plastic anchors—little sheaths that you slip into the hole—plus screws or lag bolts. For floor sills where the only possibility of movement is sideways, the only fastener needed is a spike that has been cut off short.

Powder-driven studs are the deluxe fastening device. Using them calls for owning or renting a stud driver, which works on a powder charge similar to a .320 caliber long rimfire blank cartridge. The explosion drives a hardened-steel stud into concrete so effectively that it may take a 3,000-pound pull to get it out. To fit every job, there's a choice of a dozen kinds and sizes of studs. Some studs have heads like nails, to fasten metal to masonry. Use them with disks for fastening wood or insulation board. Other types have internal or external threads, to take machine screws or nuts.

Using the stud driver is a matter of load-place-fire. Load plus stud will cost you about seventeen to twenty-three cents. You'll find it a good system for big jobs—hanging conduit clips, pipe hangers, electric boxes; fastening furring strips for wall paneling; putting down sills or floor plates on concrete slabs; fastening metal flashing to a masonry wall or chimney.

There are also mechanical stud-driving devices. They're cheaper to buy—they cost just a few dollars—than the powder-driven kind and also cheaper to use. They give support to slender steel pins, like nails, while you drive them with a hammer blow. Naturally these manual drivers won't do the work of the explosive kind, but they will often function in situations a little too tough for ordinary concrete nails.

Adhesive-held anchors are offered in several shapes, for holding furring strips, metal boxes and cabinets, insulation, conduit hangers. Basically each consists of a 2-inch square of steel to which is welded the hanging device, the square being punched full of holes. Bond is made with one of these remarkable reclaimed-rubber mastics. You smear it over the base of the hanger, then push the hanger tightly against the wall.

You may doubt the strength of such a method. I did. So I fastened a 1-by-2 furring strip to a rough concrete wall with just two anchor nails. Next day I tried to pull the strip away from the wall. I broke the furring strip in two without budging the anchor nail.

If your job is fastening up furring strips in order to give your basement a new dry wall or wood or plywood paneling, anchor nails are the right type.

Anchor bolts work the same way, but have threaded bolts instead of nails. Use them to bolt up metal cabinets, outlet boxes, salt dispensers, shelf supports, mail boxes and such things. Similarly, there are hanger supports for installing electrical conduit or cable. Pipe straps and cable clamps are made to go with them.

Insulation hangers are another anchor specialty. They go up the same way, have a wire finger over which insulation is pushed. Thus a masonry wall can be insulated as easily as a frame one.

All these adhesive-anchor devices work on practically any surface: brick, steel, concrete, gypsum tile, hollow tile, flat stone.

Adhesives will also solve some problems directly. You can use mastics to fasten furring strips and other things directly to masonry walls and to floor slabs. Panel adhesives, intended mostly for fastening plywood and other big-sheet paneling to walls, can also be used for securing furring strips. Contact-bond cement is also very effective when you are working with smooth surfaces.

TYPES OF MASONRY ANCHORS

Lead-lined, braided-jute plug is inserted in hole drilled to proper size, and fixture is then attached with screw which expands the plug. These plugs can be used in any masonry material, for attaching conduit clamps, outlet boxes, lighting fixtures, and other objects.

Toggle bolts are used in anchoring fixtures to hollow walls, whether of concrete block, gypsum board, plaster, or paneling. Toggle is pushed through hole, with fixture on bolt, and then brought up tightly against inside surface of wall by turning the bolt.

Another device for attaching fixtures to hollow walls is the expansion screw anchor. Inserted into drilled hole in the wall, it is then expanded by turning the screw. Screw is then removed, threaded through fixture hole, and replaced. Once the anchor has been securely attached, the screw may be removed and replaced as often as is necessary.

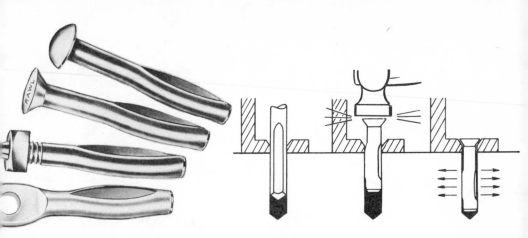

One-piece expansion bolts make attaching fixtures to masonry a quick, sure job. Bolt is inserted through fixture and into hole and merely driven home with a hammer, thereby expanding the plug in the hole. This type of expansion bolt is intended for use in hard masonry—concrete, cement, stone, extra-hard brick. It is used for such jobs as nailing wood floors on concrete, attaching machinery, acoustical ceilings, furring strips, and air-conditioner ducts. The anchors come with different heads, including a bolt-and-nut, for different types of jobs.

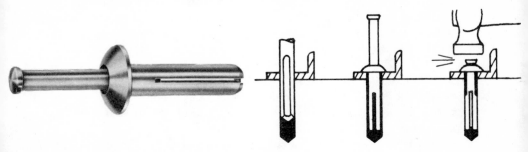

Nailin anchors are another expansion device for attaching masonry objects. Once the nail is driven home, the fixture is locked in place, leaving it unaffected by vibration. It is effective in brick, concrete, concrete block, marble, and terrazzo.

Adhesive-type anchors attach to surface of masonry wall with a black mastic adhesive. They come with nails protruding from the plate, allowing you to drive furring strips onto the nails, which are then hammered down to hold the strips firm.

CHAPTER TWENTY-SEVEN

LAY A RUBBLESTONE WALL— AND BUILD A FIREPLACE

RUBBLE IS simply uncut field or quarry stone. It is a favorite building materi-al all over the world, and it produces many of the ruggedest and most attrac-tive walls. In the hands of the tasteless user it can be responsible for some hor-rors, true, but that could equally be said of most other building materials.

Essentially there are two ways to lay up a rubble wall: with a solid form to build against, or with only your eye and perhaps some string and a level and straightedge to guide you.

Unless you're an old hand, you'll get a far more precise wall if you use some kind of solid formwork. The usual thing is lumber or plywood, used much as in building forms for poured concrete—but without having to be made so strong. For convenience in working, you'll want to build your form as you go, laying up a board at a time as you need it.

With the form boards as support for the face of the stone, you lay up one piece after another in a bed of cement mortar. Use the usual mix—one shovel-ful of portland cement to three of sand and enough water to make a plastic mortar. The sand need not be the fine kind used for plaster and most mortar; a coarser concrete sand will give you more strength and a more rugged effect.

But use fine sand for the mortar if you lay up the stone without formwork. You'll want to fit the stones somewhat more closely together in this case, shap-ing some of them to some degree with a mason's hammer for better fit.

The rubblestone fireplace wall you see in the photographs was laid up al-most entirely without the use of forms. I did the job by eye, with only an oc-casional bit of guidance from a level or a line, because all of it except the area right around the fireplace opening was in the shape of a sweeping curve. Form-ing for a curve seemed like too much work.

For fireplace building I strongly recommend the use of a warm-air-circulat-ing steel core. It makes the job far easier and guarantees you a successful, non-smoking fireplace. Where an ordinary fireplace heats you one-sidedly by radia-tion alone, the circulator adds warm air by convection. Merely useful in a house with a furnace, this feature becomes almost essential wherever a fireplace must do the whole heating job.

Basis of the fireplace unit is a double-walled steel firebox. As this heats up,

Whether it's for a fireplace or not, a rubblestone wall is strong and attractive. Such a wall is quicker, and usually far cheaper, than one laid up with cut stone.

air moves through it by gravity and flows out into the room. Topping the core is a smoke dome with hamper and, in the case of the unit I used (Heatilator, made by Vega Industries, Syracuse, N.Y. 13205) a built-in downdraft shelf.

The units are made to give these opening widths in inches: 28, 32, 36, 40, 48, 60. In choosing, consider the size of the room and the length of firewood you'll use.

Since you're dealing with a considerable mass of masonry, your fireplace will need a stout slab foundation. This may sit on a concrete footing, or be suspended above it with an ash pit between. In that case you'll want an ash dump in the slab and a cleanout door outside. Make your foundation at least as large as the area to be occupied by the masonry of the fireplace. It will have to be from 2 feet (for a small unit) to 3 feet wide and half again as long as the width of the fireplace opening. That's if you use brick or thin stonework. For other masonry, including fieldstone, figure more generously. The footing below, 8 inches to 12 inches of concrete, should extend 6 inches beyond the foundation.

To key the masonry together, embed pieces of reinforcing steel in the foundation. Use at least two ½-inch-diameter rods.

Lay up the stone for your fireplace just as if you were building any other wall. Then fill behind the stone facing with concrete plus any rock you have to spare. Let the steel core act as a form; it should not support any weight except while the concrete is setting. Bridge gaps with pieces of reinforcing steel.

151

These two photographs show early and late stages of a rubblestone fireplace wall built by the author in an addition to a large house on the California coast.

A roll of mineral wool will come with your fireplace unit. Cover all steel surfaces with this before putting masonry against them. You can hold it in place while you work by daubing the steel first with a paste of cement and water. The wool gives a cushion for expansion which otherwise might produce cracking when you have a fire.

You must provide ducts from the outside of the wall to the two intakes and two outlets. The intakes should be as close to the floor as possible.

Since the fireplace shown is standing free near the center of a large room, its air intakes could be put at the back. Its outlets are at the ends of the wall, for looks and for heating a large room.

Where ducts are short, they are easily made in the masonry by using wood forms. For very long ones, use ordinary galvanized heating ducts of the kind that roll up and snap together like stovepipe.

Use steel angle to reinforce and support the top of the fireplace opening. A piece of the correct size will be available where you purchase the unit. Use lumber to make a temporary form at this point, setting the stones in a bed of reinforced concrete several inches thick. Pack mineral wool between the angle-steel lintel and the unit.

After you reach the top of the steel unit you may wish to continue the stone structure full size up to the ceiling or even to full chimney height. Or you may narrow it to normal chimney size, either at the top of the unit or at ceiling level. In any case, the first step is to place a length of terra-cotta flue liner of the size specified in the instruction sheet that comes with all fireplace cores. The flue liner must be supported by masonry, not by the steel unit. To make sure of this, carry your masonry above the top of the steel and set the liner on it. Continue upward, adding liner as you go. You should have at least 4 inches of concrete around the liner, reinforced with a vertical steel rod at each corner.

Continue the chimney until it is at least 3 feet higher than the roof where it passes through. It should also be at least 2 feet higher than any part of the roof within 10 feet.

Finish by sloping the cement upward from the face of the chimney toward the liner. Let the flue extend about 2 inches above the rest of the chimney. These are precautions to discourage downdrafts.

1. Having cast concrete footing and hearth, move the circulator unit into place. Pipe cap at center of base is end of line for a gas kindler.

2. Paste mineral wool to unit so that no masonry comes into contact with steel of the core. This will provide space for expansion and prevent cracking later.

3. For a rugged wall you can use fieldstone pretty much as it comes if you choose each piece carefully for fit. Mason's hammer is useful for occasional shaping.

4. You'll need simple formwork like this to support concrete and angle-steel lintel over opening while concrete sets. Steel-reinforce masonry over all openings.

5. Surround reinforcing with 2 inches or more of concrete over all openings, and set stones into this. Wirebrushing this concrete makes it blend better with stones and mortar.

6. A fireplace like this is basically a thick rubblestone wall. Its center is made of ordinary concrete shoveled in. Steel bars stubbed up give added strength.

7. Top of the steel dome must be split at corners and bent to fit flue liner. The cardboard template seen lying here permits checking the fit without lifting the heavy liner.

8. To build a woodbox, or to create openings for any purpose in a rubble-stone wall, use lumber or plywood forms on the sides and at the top.

9. Ducts from the fireplace core bring heated air to the ends of the wall for better room heating. Ducts at floor level are also needed, for cold-air intake.

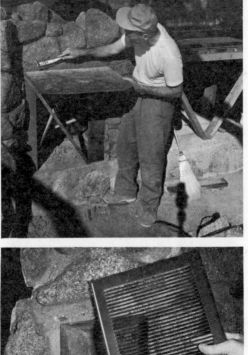

CHAPTER TWENTY-EIGHT

POUR A STONE-IN-CONCRETE WALL—AND A FIREPLACE TOO

ONE OF the many things the brilliant American architect Frank Lloyd Wright taught us is that magnificent masonry walls and fireplaces can be built by unskilled hands. Wright's famous Phoenix headquarters, Taliesin West, is a riot of colorful desert rocks nested in gray concrete. Since the technique calls for simply placing flat-faced stones against the forms while pouring concrete, none of the usual stonemason's skills are called for.

To build a stone wall of this kind, you erect forms much like those for pouring a wall of ordinary concrete. The only difference is that you may stop the boards lining the form at a height of no more than 3 feet.

When you pour the wall, stop after reaching a height of about ground level. Place large flat stones inside the forms, with their flat faces against the boards. If necessary, wedge the stones there to keep them from tipping while you continue the concrete pour. As each stone is covered, place others similarly. But don't place them in level courses or any consistent pattern. A random effect of large and small stones is what you want. Continue on up the wall this way, building up the form and inserting more rocks as you go.

Pull the forms off as soon as your wall is strong enough to support itself safely. It is better if this is not later than the next morning. If you do this, you won't have to use too much effort in order to scrape and chisel away concrete that has spilled down over rock faces that you want exposed.

There is another touch to which I am partial, though I don't know whether Mr. Wright would have approved. Give the surface of the concrete the same treatment that produces an exposed-aggregate slab, described in an earlier chapter. To do this, pull the forms off very early, usually within four hours, proceeding cautiously until you are certain the wall will support itself.

Then hose down the wall to remove all the cement and fine sand near the surface and permit the gravel in the aggregate to show. You can spray water violently and make quite a rough wall if you wish, much rougher than you would care to have underfoot. The effect you will get will depend upon how soon you spray and how hard—and also upon what kind of aggregate is contained in the concrete. In ordering the concrete, or the aggregate from which

156

to make it, discuss the intended use with the supplier. He may be able to give you something especially suitable.

This hosing process also does an excellent job of cleaning the big rocks and you can use the water from the hose to knock off any chunks of concrete that may be covering them.

Unfortunately all this is a pretty messy operation, involving a large volume of water loaded with sand and cement. So it won't be feasible to use this method in building a fireplace or planter wall inside an existing house. But you can still do an ordinary rock-in-concrete job.

Either way, a wall of this kind fits into a surprising number of places. Its rough hominess adapts it perfectly for all garden and yard use. One reason it blends in so well is that it looks like the product of a hundred years of wind and weather the day you build it. In my own home, poured-and-hosed masonry of this kind makes up all the visible walls of the first story, as well as the fireplace-and-barbecue wall shown in the pictures.

By combining your poured-concrete-and-rocks with a steel fireplace core, such as the Heatilator described in the preceding chapter, you can pour a fireplace. This is a far quicker method than laying up rubble, just as the rubble system is a simplification by comparison with traditional cut-stone methods. The choice between the two methods will depend in large part upon the type of stone available. Good-sized stones with relatively flat faces are essential for the poured method.

For a striking poured wall, use as large stones as you can handle. You may

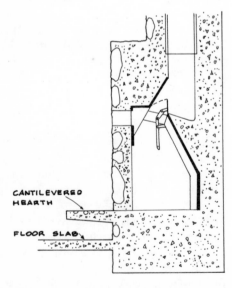

Drawing shows how the circulator-fireplace core acts as inside form for a poured-concrete fireplace. Colorful flat-faced rocks have been placed against the wooden outside forms at the front of the fireplace. Rear wall can be done in the same fashion if it will show. The cantilevered hearth has a surface of pebble concrete made as described in Chapter 9.

even wish to do what I did, in the absence of the corps of husky apprentices that Wright had available: use a rope hoist to lift such big chunks of stone as the one right over the center of my fireplace opening.

Liberal use of steel reinforcing will make your wall tremendously strong. By using plenty of ordinary ½-inch rod, I made mine so stout that the building inspector unhesitatingly accepted it for structural purposes—as support of the main beam and upper wall of my two-story house. This is not permitted under any building code I know of with ordinary masonry fireplace construction.

In the upper part of your fireplace you'll find places where the walls are far thicker than necessary. You can use this space (and save concrete) by casting in niches or cupboards, forming them with boards.

You can make what is sometimes called a Swedish fireplace by casting the base and hearth to a level about 16 inches higher than the floor of the room. Such a fireplace is easier to fuel and to cook over, and it throws its heat more effectively. Remember that cool-air intakes should be as close to floor level as possible and that if your hearth is cantilevered it should be heavily reinforced with steel rods running both ways. The pebble finish described in an earlier chapter makes a very attractive hearth.

A gas kindler will add enormously to the utility of your fireplace at a cost of only a few dollars. Various types are available from dealers.

The general principles and main steps to follow in pouring a fireplace of your own are given below. Details such as working dimensions and measurements and flue sizes and chimney heights naturally vary with the size of the unit which you choose. These figures can be found on the instruction sheets that come with the circulator units.

1. Excavate at least 8 inches below ground level, or to frost line if this is deeper, and pour a concrete foundation slab and hearth. Make the hearth generous—at least 24 inches from front to back and 24 inches wider than the fireplace opening. Plug in lengths of ½ inch reinforcing rqd where the wall around the unit is to go.

2. Move the circulator unit into place on the slab. Set it in thin cement mortar.

3. With scraps of brick, stone, sheet metal, or asbestos board, build a duct from each of the air intakes to where you will place a grille. These passageways can be crude since they will be buried in the concrete you are about to pour.

4. Build a form all the way around the unit, at least 8 inches from it at all points. A good choice for form lumber is 2-by-6 tongue-and-groove subflooring, fastened with duplex nails for easy tearing-down and salvage. Continue the form only to top of the front opening in the unit. Extend the two sides of this opening from the unit to the form boards. Pad the surface of the unit with mineral wool so that no concrete will touch the steel.

5. Pour concrete to the height of your form work. Add reinforcing steel as you go, and place flat-faced rocks against the forms where they will be exposed when the forms are pulled.

6. Continue the form on up in the same way for the second and succeeding pours. Mortar in small flat rocks to form a series of channels from the warm-air outlets in the unit.

Inserted rocks and exposed aggregate give this wall color and texture without the spottiness found in so much stonework.

7. As you cast concrete at the top of the unit, shape it to support the first of the terra cotta flues that will line the chimney.

8. With flue liner sections as the inside form and lumber or plywood as the outside, pour a chimney in much the same way as you have done the fireplace. For good appearance—and probably far more strength than necessary—make walls 8 inches thick or more. Reinforce with at least one piece of steel at each corner.

9. Continue the chimney to a point at least 2 feet above the highest point of the roof—3 feet with a flat or low-pitched roof. Extra height improves draft. Let the last piece of flue liner extend several inches higher than the concrete you pour around it.

10. Waterproof the connection between chimney and roof with flashing. Add a chimney rain cap if you want one and a spark guard if required by local fire regulations or conditions.

Closeup shows texture obtained in concrete surrounding the rocks by hosing strongly and scrubbing with brush where necessary.

A few really big rocks greatly improve the effect. Hoist them into the wall, build the form, then lean them against it.

Lintel for any opening in a cast wall calls for formwork like this, supporting top of fireplace opening and cantilever hearth.

Opening at left is for barbecue, with storage cupboard beneath it. Next opening is wood-box. Cupboard form is above it.

Steel barbecue unit sits in cast opening, will have concrete walls cast at sides and back, with damper and flue above.

Rubber-gloved hands are often the best tool for working with mortar. Openings being made between rocks are for warm-air outlet.

Small forms, as for this cast-concrete chimney, are quickly made by nailing plywood panels together, then reinforcing the nails with clamps.

Strong chimney is reinforced cast concrete with flue liner. Correct flue size is given in the instructions which come with each fireplace unit.

To create the best draft, the top of the chimney should slope slightly downward from the flue liner, which should protrude a couple of inches.

CHAPTER TWENTY-NINE

SEVEN BARBECUES
YOU CAN BUILD

ANYONE WHO has lived on a hill above a residential suburb and sniffed the air on a balmy evening knows how much a part of American life today the outdoor barbecue is. The outdoor cooking unit can be as simple as the firepit you see in one of the photographs here—or as elaborate as the grill-and-fireplace combination shown in another.

A masonry niche into which fits a purchased steel unit is the simplest barbecue grille of them all. A poured-concrete variation on this is shown in the preceding chapter. Note the attractive use here of screen block and the pebble-concrete floor cast in a nicely spaced grid of 2-by-4 California redwood.

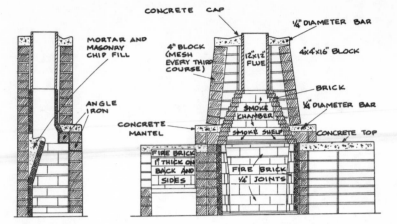

CONCRETE CAP

MORTAR AND MASONRY CHIP FILL

¼" DIAMETER BAR

4" BLOCK (MESH EVERY THIRD COURSE)

12"x12" FLUE

4"x4"x16" BLOCK

BRICK

ANGLE IRON

SMOKE CHAMBER

¼" DIAMETER BAR

CONCRETE MANTEL

SMOKE SHELF

CONCRETE TOP

FIRE BRICK 1" THICK ON BACK AND SIDES

FIRE BRICK ⅛" JOINTS

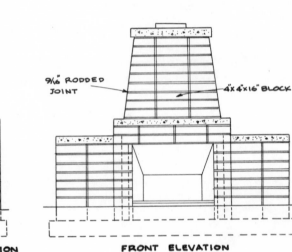

9/16" RODDED JOINT

4"x4"x16" BLOCK

LEFT ELEVATION

FRONT ELEVATION

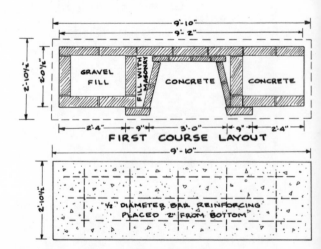

9'-10"

9'-2"

2'-10½"

2'-0½"

GRAVEL FILL

FILL WITH MASONRY

CONCRETE

CONCRETE

2'-4"   9"   3'-0"   9"   2'-4"

FIRST COURSE LAYOUT

9'-10"

2'-10½"

½" DIAMETER BAR REINFORCING PLACED 2" FROM BOTTOM

MAKE DEEP ENOUGH TO EXTEND BELOW LOCAL FROST LEVEL

FOOTING

These diagrams show construction of the fireplace-barbecue combination you see in the third photograph. Although some quantities will vary, depending upon decisions made as you go along, depth of footing required, and so on, the following list of materials should give you a pretty good idea of what you'll need:

Cement, sand, and gravel to make concrete for footing
Masonry cement and sand for mortar, or dry-mix mortar
Two ⅜-inch angle irons, 3''x3''x39''
Metal grill unit
Sack of fire-clay mortar
Reinforcing rod and mesh
20 concrete partition blocks
78 fire brick
90 fire-resistant concrete brick
213 4''x4''x16'' concrete blocks, which can be split block
Flue liner 12''x12''x24''

A firepit's no great cooking device, but it will serve—and it makes a fine people-warming outdoor fireplace as well. Two hints: provide a stout cover against rain and accidents in periods of nonuse; add a gas-fired kindler (note control device between pit and chair) and you'll use the pit far more.

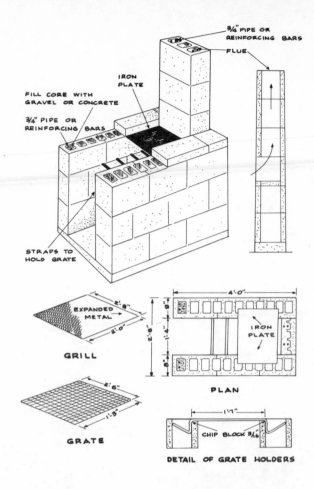

FILL CORE WITH GRAVEL OR CONCRETE

¾" PIPE OR REINFORCING BARS

IRON PLATE

¾" PIPE OR REINFORCING BARS

FLUE

STRAPS TO HOLD GRATE

EXPANDED METAL

2'-8"

2'-0"

GRILL

2'-6"

1'-3"

GRATE

4'-0"

IRON PLATE

2'-8"

1'-4"

8"

8"

PLAN

1'-7"

CHIP BLOCK 3/16

DETAIL OF GRATE HOLDERS

Building this efficient backyard barbecue is a simple and straightforward procedure. Just assemble and mortar together the block as shown, on a base of poured concrete 4 to 6 inches thick. You will have to work out your own method of forming the smoke shelf and throat, depending upon the type of block available where you live. The grate holders can be strap steel about 3/16-by-1-inch, or you can use reinforcing rod. Either way, there should be five of them. The grate rests directly on these and can be slid forward or back for control of the fire. The grill, of expanded metal lath, will rest on the top of the blocks. For strength, reinforce at least the cores at the four corners and the two at the sides of the flue. Use one or two lengths of reinforcing rod or old pipe in each, and pour the core full of gravel or—preferably—concrete. One attractive variation on the design shown here is the use of block 4 inches high instead of 8 inches, in six courses instead of three. Whether this is regular block or the split or the slump type, the brick shape usually harmonizes better with yard or garden surroundings.

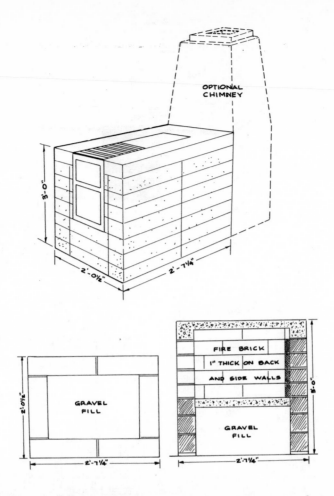

OPTIONAL CHIMNEY

3'-0"

2'-0½"

2'-7¼"

GRAVEL FILL

2'-0½"

2'-7¼"

FIRE BRICK
1" THICK ON BACK
AND SIDE WALLS

GRAVEL FILL

3'-0"

2'-7¼"

This is the quickest of all outdoor barbecues to build since it uses a manufactured metal grill unit. All you build is a concrete-block housing for this grill, placing this on a base of poured concrete. Add a chimney if you feel like it. Details and dimensions will vary since they are dependent upon the size of the metal unit, but the sketches will give you a pretty good idea of how to go about constructing a barbecue of this kind. The details here assume that you are using 4-by-4-by-16-inch concrete block (the split kind looks better than the ordinary), but you can easily adapt the idea to blocks of any other size.

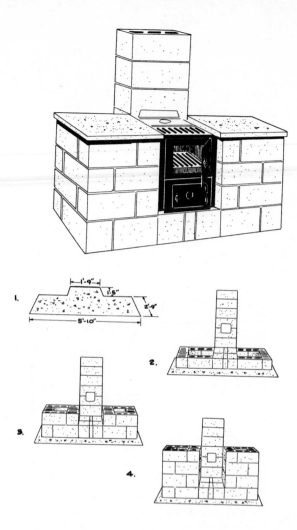

Here's another barbecue that uses a purchased steel unit in a concrete-block base. Dimensions will naturally vary with the unit you choose and the kind of block available. Size of base and arrangement of blocks sketched here assumes the use of 17-by-21-inch chimney blocks. The other blocks are 8-by-8-by-16-inch units. As with the other barbecues, this one can be made more attractive—to my eye, at least—by substituting twice as many blocks 4 inches high. Best procedure in casting a base for a barbecue is to begin by digging to a depth below frost. (In a warm climate 6 inches is enough.) Having done this, fill with sand and gravel until about 6 inches below ground level. Pour the base slab, reinforcing it with wire mesh. While you're about it, make the two concrete cap slabs; for this combination of block and unit, they are 25-by-32½ inches. They should be 1 inch thick and also reinforced. After the base has cured for at least two days you can lay up the block. Use hammer and chisel to cut a 5-inch-square smoke inlet opening into the flue close under the top edge of the metal fireplace unit.

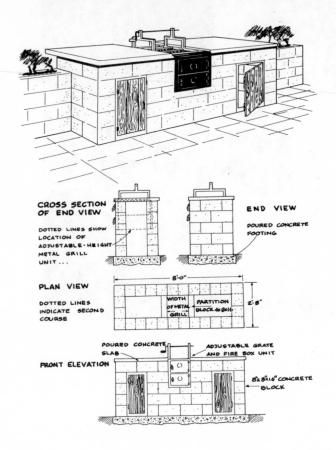

Generous storage space, generous counter space, and a metal grill unit that includes a spit are the features of this hospitable barbecue. It can be built into a wall, as you see it here, or made freestanding—or attached to an outdoor fireplace. Unless you have other blocks as part of your house or garden structures or walls and want to match them, you should consider using half-height blocks instead of the 8-inch ones shown here for simplicity. The smaller blocks give the more pleasing shape made familiar by bricks and adobe. As designed and built for the National Concrete Masonry Association, this backyard kitchen has a metal grill unit 21-by-26-by-15½-inches. The two counter-tops are pieces of precast concrete 2-by-33-by-35 inches. These should be well cured before being lifted into place—an operation that will call for some neighborly assistance. Along with sand and gravel for the concrete, sand and lime (optional) for the mortar, and cement for both, you'll need a pair of doors, which can be ¾-by-16-by-32-inch exterior-grade plywood, and concrete blocks. Your list of these will vary with the size and style you use, but in one type of 8-by-8-by-16-inch block the following are called for: 20 single-corner; 39 standard; 14 partition. These last are 4-by-8-by-16 inches. For convenience in case of variation in design or loss by breakage, it is always well to buy a few extra blocks. They're returnable.

This impressive triple-threat is an outdoor fireplace flanked by a barbecue and a storage cupboard. The diagrams show the essentials of the construction method used, whether you wish to follow this closely or merely take it as a point of departure. Principal material is 4-by-4-by-16-inch regular or split concrete block.

Construction sketches show how to build all but the two simplest of the seven barbecues in this chapter. Whether you build with concrete block, as the sketches suggest, or adapt the plans to other masonry materials, all the how-to you should need can be found in earlier chapters.

You can, of course, vary the styling immensely by your choice of materials. Using split block instead of the ordinary kind will in itself give your barbecue unit a trimmer and more professional look. Slump block will make it look cozier and more homespun. And merely substituting 4-inch-high block for the more common, and more commonplace, 8-inch type will be a distinct improvement.

By using techniques described and shown in the two chapters preceding this one, you can build a stone barbecue to any of the plans sketched. Be guided both by what materials are most readily available locally and by what will match or harmonize best with your house and especially any masonry parts it may have—facing, chimney, and so on.

You can use many of the other ideas and methods in earlier chapters to make your barbecue area more attractive and more useful. A masonry wall that links the unit to the house is one possibility. Paving of at least the cooking area is almost essential. Extending the paving to create an outdoor dining—and perhaps lounging—area is even better.

INDEX